Tucholsky Wagner Zola Scott Sydow Schlegel
Turgenev Wallace Fonatne Freud
Twain Walther von der Vogelweide Fouqué Friedrich II. von Preußen
Weber Freiligrath Frey
Fechner Fichte Weiße Rose von Fallersleben Kant Ernst Richthofen Frommel
Engels Fielding Hölderlin
Fehrs Faber Flaubert Eichendorff Tacitus Dumas
Maximilian I. von Habsburg Eliasberg Ebner Eschenbach
Feuerbach Fock Eliot Zweig
Ewald Vergil
Goethe London
Mendelssohn Balzac Shakespeare Elisabeth von Österreich Dostojewski Ganghofer
Trackl Lichtenberg Rathenau Doyle Gjellerup
Stevenson Hambruch
Mommsen Tolstoi Lenz Hanrieder Droste-Hülshoff
Thoma von Arnim
Dach Verne Hägele Hauff Humboldt
Karrillon Reuter Rousseau Hagen Hauptmann Gautier
Garschin Defoe Hebbel Baudelaire
Damaschke Descartes Hegel Kussmaul Herder
Wolfram von Eschenbach Dickens Schopenhauer Rilke George
Darwin Melville Grimm Jerome Bebel Proust
Bronner Campe Horváth Aristoteles Barlach Voltaire Federer Herodot
Bismarck Vigny Gengenbach Heine
Storm Casanova Tersteegen Gilm Grillparzer Georgy
Chamberlain Lessing Langbein Gryphius
Brentano Claudius Schiller Lafontaine
Strachwitz Bellamy Schilling Kralik Iffland Sokrates
Katharina II. von Rußland Gerstäcker Raabe Gibbon Tschechow
Löns Hesse Hoffmann Gogol Wilde Gleim Vulpius
Luther Heym Hofmannsthal Klee Hölty Morgenstern Goedicke
Roth Heyse Klopstock Kleist
Luxemburg Puschkin Homer Möricke
La Roche Horaz Musil
Machiavelli Kierkegaard Kraft Kraus
Navarra Aurel Musset Lamprecht Kind Kirchhoff Hugo Moltke
Nestroy Marie de France
Nietzsche Nansen Laotse Ipsen Liebknecht
Marx Lassalle Gorki Klett Ringelnatz
von Ossietzky May vom Stein Lawrence Leibniz
Petalozzi Platon Puschkin Knigge Irving
Sachs Pückler Michelangelo Kafka
Poe Liebermann Kock
de Sade Praetorius Mistral Zetkin Korolenko

The publishing house tredition has created the series **TREDITION CLASSICS**. It contains classical literature works from over two thousand years. Most of these titles have been out of print and off the bookstore shelves for decades.

The book series is intended to preserve the cultural legacy and to promote the timeless works of classical literature. As a reader of a **TREDITION CLASSICS** book, the reader supports the mission to save many of the amazing works of world literature from oblivion.

The symbol of **TREDITION CLASSICS** is Johannes Gutenberg (1400 – 1468), the inventor of movable type printing.

With the series, tredition intends to make thousands of international literature classics available in printed format again – worldwide.

All books are available at book retailers worldwide in paperback and in hardcover. For more information please visit: www.tredition.com

tredition was established in 2006 by Sandra Latusseck and Soenke Schulz. Based in Hamburg, Germany, tredition offers publishing solutions to authors and publishing houses, combined with worldwide distribution of printed and digital book content. tredition is uniquely positioned to enable authors and publishing houses to create books on their own terms and without conventional manufacturing risks.

For more information please visit: www.tredition.com

Steam, Steel and Electricity

James W. Steele

Imprint

This book is part of the TREDITION CLASSICS series.

Author: James W. Steele
Cover design: toepferschumann, Berlin (Germany)

Publisher: tredition GmbH, Hamburg (Germany)
ISBN: 978-3-8491-5099-0

www.tredition.com
www.tredition.de

Copyright:
The content of this book is sourced from the public domain.

The intention of the TREDITION CLASSICS series is to make world literature in the public domain available in printed format. Literary enthusiasts and organizations worldwide have scanned and digitally edited the original texts. tredition has subsequently formatted and redesigned the content into a modern reading layout. Therefore, we cannot guarantee the exact reproduction of the original format of a particular historic edition. Please also note that no modifications have been made to the spelling, therefore it may differ from the orthography used today.

STEAM STEEL AND ELECTRICITY

By

JAMES W. STEELE

CONTENTS

THE STORY OF STEAM.

What Steam is.—Steam in Nature.—The Engine in its earlier forms.—Gradual explosion.—The Hero engine.—The Temple-door machine.—Ideas of the Middle Ages.—Beginnings of the modern engine.—Branca's engine.—Savery's engine.—The Papin engine using cylinder and piston.—Watt's improvements upon the Newcomen idea.—The crank movement.—The first use of steam expansively.—The "Governor."—First engine by an American Inventor.—Its effect upon progress in the United States.—Simplicity and cheapness of the modern engine.—Actual construction of the modern engine.—Valves, piston, etc., with diagrams.

THE AGE OF STEEL.

The various "Ages" in civilization.—Ancient knowledge of the metals.—The invention and use of Bronze.—What Steel is.—The "Lost Arts."—Metallurgy and chemistry.—Oriental Steel.—Modern definition of Steel.—Invention of Cast Steel.—First iron-ore discoveries in America.—First American Iron-works.—Early methods without steam.—First American casting.—Effect of iron industry upon independence.—Water-power.—The trip-hammer.—The steam-hammer of Nasmyth.—Machine-tools and their effects.—First rolling-mill.—Product of the iron industry in 1840-50.—The modern nail, and how it came.—Effect of iron upon architecture.—The "Sky-Scraper."—Gas as fuel in iron manufactures.—The Steel of the present.—The invention of Kelley.—The Bessemer process.—The "Converter."—Present product of Steel.—The Steel-mill.

THE STORY OF ELECTRICITY.

The oldest and the youngest of the sciences.—Origin of the name.—Ancient ideas of Electricity.—Later experiments.—Crude notions and wrong conclusions.—First Electric Machine.—Frictional Electricity.—The Leyden Jar.—Extreme ideas and Fakerism.—Franklin, his new ideas and their reception.—Franklin's Kite.—The Man Franklin.—Experiments after Franklin, leading to our present modern uses.—Galvani and his discovery.—Volta, and the first "Battery."—How a battery acts.—The laws of Electricity, and how they were discovered.—Induction, and its discoverer.—The line at which modern Electricity begins.—Magnetism and Electricity.—The Electro-Magnet.—The Molecular theory.—Faraday, and his Law of Magnetic Force.

MODERN ELECTRICITY.

CHAPTER I.

The Four great qualities of Electricity which make its modern uses possible.—The universal wire.—Conductors and non conductors.—Electricity an exception in the ordinary Laws of Nature.—A dual nature: "Positive" and "Negative."—All modern uses come under the law of Induction.—Some of the laws of this induction.—Magnets and Magnetism.—Relationship between the two.—Magnetic "poles."—Practical explanation of the action of induction.—The Induction Coil.—Dynamic and Static Electricity.—The Electric Telegraph.—First attempts.—Morse, and his beginnings.—The first Telegraph Line.—Vail, and the invention of the dot-and-dash alphabet.—The old instruments and the new.—The final simplicity of the telegraph.

CHAPTER II.

The Ocean Cable. — Differences between land lines and cables. — The story of the first cable. — Field and his final success. — The Telephone. — Early attempts. — Description of Bell's invention. — The Telautograph. — Early attempts and the idea upon which they were based. — Description of Gray's invention. — How a Telautograph may be made mechanically.

CHAPTER III.

The Electric Light. — Causes of heat and light in the conductor of a current. — The first Electric Light. — The Arc Light, and how constructed. — The Incandescent. — The Dynamo. — Date of the invention. — Successive steps. — Faraday the discoverer of its principle. — Pixü's machine. — Pacinatti. — Wilde. — Siemens' and Wheatstone. — The Motor. — How the Dynamo and Motor came to be coupled. — Review of first attempts. — Kidder's battery. — Page's machine. — Electric Railroads. — Electrolysis. — General facts. — Electrical Measurements. — "Death Current." — Instruments of Measurement. — Electricity as an Industry. — Medical Electricity. — Incomplete possibilities. — What the "Storage Battery" is.

CHAPTER IV.

Electrical Invention in the United States. — Review of the careers of Franklin, Morse, Field, Edison and others. — Some of the surprising applications of Electricity. — The Range-Finder. — Cooking and heating by Electricity.

THE STORY OF STEAM

That which was utterly unknown to the most splendid civilizations of the past is in our time the chief power of civilization, daily engaged in making that history of a new era that is yet to be written in words. It has been demonstrated long since that men's lives are to be influenced not by theory, or belief, or argument and reason, so much as by that course of daily life which is not attempted to be governed by argument and reason, but by great physical facts like steam, electricity and machinery in their present applications.

The greatest of these facts of the present civilization are expressed in the phrase, Steam and Steel. The theme is stupendous. Only the most prominent of its facts can be given in small space, and those only in outline. The subject is also old, yet to every boy it must be told again, and the most ordinary intelligence must have some desire to know the secrets, if such they are, of that which is unquestionably the greatest force that ever yielded to the audacity of humanity. It is now of little avail to know that all the records that men revere, all the great epics of the world, were written in the absence of the characteristic forces of modern life. A thousand generations had lived and died, an immense volume of history had been enacted, the heroes of all the ages, and almost those of our own time, had fulfilled their destinies and passed away, before it came about that a mere physical fact should fill a larger place in our lives than all examples, and that the evanescent vapor which we call steam should change daily, and effectively, the courses and modes of human action, and erect life upon another plane.

It may seem not a little absurd to inquire now "what is steam?" Everybody knows the answer. The non-technical reader knows that it is that vapor which, for instance, pervades the kitchen, which issues from every cooking vessel and waste-pipe, and is always white and visible, and moist and warm. We may best understand an answer to the question, perhaps, by remembering that steam is one of the three natural conditions of water: ice, fluid water, and steam.

One or the other of these conditions always exists, and always under two others: pressure and heat. When the air around water reaches the temperature of thirty-two degrees by the scale of Fahrenheit, or ° or zero by the Centigrade scale, and is exposed to this temperature for a time, it becomes ice. At two hundred and twelve degrees Fahrenheit it becomes steam. Between these two temperatures it is water. But the change to steam which is so rapid and visible at the temperature above mentioned is taking place slowly all the time when water, in any situation, is exposed to the air. As the temperature rises the change becomes more rapid. The steam-making of the arts is merely that of all nature, hastened artificially and intentionally.

The element of pressure, mentioned above, enters into the proposition because water boils at a lower temperature, with less heat, when the weight of the atmosphere is less than normal, as it is at great elevations, and on days when, as we now express it, there is a low barometer. Long before any cook could explain the fact it was known that the water boiling quickly was a sign of storm. It has often been found by camping-parties on mountains that in an attempt to boil potatoes in a pot the water would all "boil away," and leave the vegetables uncooked. The heat required to evaporate it at the elevation was less than that required to cook in boiling water. It is one of the instances where the problems of nature intrude themselves prominently into the affairs of common life without previous notice.

This universal evaporation, under varying circumstances, is probably the most important agency in nature, and the most continuous and potent. There was only so much water to begin with. There will never be any less or any more. The saltness of the sea never varies, because the loss by evaporation and the new supply through condensation of the steam—rain—necessarily remain balanced by law forever. The surface of our world is water in the proportion of three to one. The extent of nature's steam-making, silent, and mostly invisible, is immeasurable and remains an undetermined quantity. The three forms of water combine and work together as though through intentional partnership, and have, thus combined, already changed the entire land surface of the world from what it was to what it is, and working ceaselessly through endless cycles will change it yet more. The exhalations that are steam become the water in a rock-cleft. It changes to ice with a force almost beyond measurement in the orderly arrangement of its crystals in compliance with an immutable law for such arrangement, and rends the rock. The process goes on. There is no high mountain in any land where water will not freeze. The water of rain and snow carries away the powdered remains from year to year, and from age to age. The comminuted ruins of mountains have made the plains and filled up and choked the mouth of the Mississippi. The soil that once lay hundreds of miles away has made the delta of every river that flows into the sea. The endless and resistless process goes on without ceasing, a force that is never expended, and but once interrupted within the knowledge of men, then covered a large area of the world with a sea of ice that buried for ages every living thing.

The common idea of the steam that we make by boiling water is that it is all water, composed of that and nothing else, and this conception is gathered from apparent fact. Yet it is not entirely true. Steam is an invisible vapor in every boiler, and does not become what we know by sight as steam until it has become partly cooled. As actual steam uncooled, it is a gas, obeying all the laws of the permanent gases. The creature of temperature and pressure, it changes from this gaseous form when their conditions are removed, and in the change becomes visible to us. Its elasticity, its power of yielding to compression, are enormous, and it gives back this elasticity of compression with almost inconceivable readiness and swiftness. To the eye, in watching the gliding and noiseless movements of one of the great modern engines, the power of which one has only a vague and inadequate conception seems not only inexplicable, but gentle. The ponderous iron pieces seem to weigh nothing. There is a feeling that one might hinder the movement as he would that of a watch. There is an inability to realize the fact that one of the mightiest forces of nature is there embodied in an easy, gliding, noiseless impulse. Yet it is one that would push aside massy tons of dead weight, that would almost unimpeded crush a hole through the enclosing wall, that whirls upon the rails the drivers of a locomotive weighing sixty tons as though there were no weight above them, no bite upon the rails. There is an enormous concentration of force somewhere; of a force which perhaps no man can fairly estimate; and it is under the thin shell we call a boiler. Were it not elastic it could not be so imprisoned, and when it rebels, when this thin shell is torn like paper, there is a havoc by which we may at last inadequately measure the power of steam.

We have in modern times applied the word "engine" almost exclusively to the machine which is moved by the pressure of steam. Yet we might go further, since one of the first examples of a pressure engine, older than the steam machine by nearly four hundred years, is the gun. Reduced to its principle this is an engine whose operation depends upon the expansion of gas in a cylinder, the piston being a projectile. The same principle applies in all the machines we know as "engines." An air-engine works through the expansion of air in a cylinder by heat. A gas-engine, now of common use, by the expansion, which is explosion, caused by burning a mixture of coal-gas and air, and the steam-engine, the universal power generator of modern life, works by the expansion of the vapor of water as it is generated by heat. Steam may be considered a species of *gradual* explosion applied to the uses of industry. It often becomes a real one, complying with all the conditions, and as destructive as dynamite.

It cannot be certainly known how long men have experimented with the expansive force of steam. The first feeble attempt to purloin the power of the geyser was probably by Hero, of Alexandria, about a hundred and thirty years before Christ. His machine was also the first known illustration of what is now called the "turbine" principle; the principle of *reaction* in mechanics. [Footnote: This principle is often a puzzle to students. There is an old story of the man who put a bellows in his boat to make wind against the sail, and the wind did not affect the sail, but the boat went backward in an opposite direction from the nozzle of the bellows. There is probably no better illustration of reaction than the "kick" of a gun, which most persons know about. The recoil of a six-pound field piece is usually from six to twelve feet. It can be understood by supposing a gun to be loaded with powder and an iron rod longer than the barrel to be left on the charge. If the outer end of this rod were then placed against a tree, and the gun were fired, it is manifest that the gun would become the projectile, and be fired off of the rod backward or burst. In ordinary cases the air in the bore, and immediately outside of the muzzle, acts comparatively, and in a measure, as the supposed rod against the tree would. It gives way, and is elastic, but not as quickly as the force of the explosion acts, and the gun is pushed backwards. It is the turbine principle, running into hundreds of uses in mechanics.] He made a closed vessel from whose opposite sides radiated two hollow arms with holes in their sides, the holes being on opposite sides of the tubes from each other. This vessel he mounted on an upright spindle, and put water in it and heated the water. The steam issuing from the holes in the arms drove them backward. The principle of the action of Hero's machine has been accepted for two thousand years, though never in a steam-engine. It exists under all circumstances similar to his. In water, in the turbine wheel, it has been made most efficacious. The power applied now for the harnessing of Niagara for the purpose of sending electric currents hundreds of miles is the turbine wheel.

[Illustration: THE SUPPOSED HERO ENGINE.]

Hero appears to the popular imagination as the greatest inventor of the past. Every school boy knows him. Archimedes, the Greek, was the greater, and a hundred and fifty years the earlier, and was the author of the significance of the word "Eureka," as we use it now. But Hero was the pioneer in steam. He made the first steam-engine, and is immortal through a toy.

The first *practical* device in which expansion was used seems to have been for the exploiting of an ecclesiastical trick intended to impress the populace. There is a saying by an antique wit that no two priests or augurs could ever meet and look at each other without a knowing wink of recognition. Hero is said to have been the author of this contrivance also. The temple doors would open by themselves when the fire burned on the altar, and would close again when that fire was extinguished, and the worshippers would think it a miracle. It is interesting because it contained the principle upon which was afterwards attempted to be made the first working low-pressure or atmospheric steam-engine. Yet it was not steam, but air, that was used. A hollow altar containing air was heated by the fire being kindled upon it. The air expanded and passed through a pipe into a vessel below containing water. It pressed the water out through another pipe into a bucket which, being thereby made heavier, pulled open the temple doors. When the fire went out again there was a partial vacuum in the vessel that had held the water at first, and the water was sucked back through the pipe out of the bucket. That became lighter again and allowed the doors to close with a counter-weight. All that was then necessary to convince the populace of the genuineness of the seeming miracle was to keep them from understanding it. The machinery was under the floor. There have been thousands of miracles since then performed by natural agencies, and there have passed many ages since Hero's machine during which not to understand a thing was to believe it to be supernatural.

[Illustration: THE TEMPLE-DOOR TRICK.]

From the time of Hero until the seventeenth century there is no record of any attempt being made to utilize steam-pressure for a practical purpose. The fact seems strange only because steam-power is so prominent a fact with ourselves. The ages that intervened were, as a whole, times of the densest superstition. The human mind was active, but it was entirely occupied with miracle and semi-miracle; in astrology, magic and alchemy; in trying to find the key to the supernatural. Every thinker, every educated man, every man who knew more than the rest, was bent upon finding this key for himself, so that he might use it for his own advantage. During all those ages there was no idea of the natural sciences. The key they lacked, and never found, that would have opened all, is the fact that in the realm of science and experiment there is no supernatural, and only eternal law; that cause produces its effect invariably. Even Kepler, the discoverer of the three great laws that stand as the foundation of the Copernican system of the universe, was in his investigations under the influence of astrological and cabalistic superstitions. [Footnote: Kepler, a German, lived between 1571 and 1630. His life was full of vicissitudes, in the midst of which he performed an astonishing Even the science of amount of intellectual labor, with lasting results. He was the personal friend of Galileo and Tycho Brahe, and his life may be said to have been spent in finding the abstract intelligible reason for the actual disposition of the solar system, in which physical cause should take the place of arbitrary hypothesis. He did this.] medicine was, during those ages, a magical art, and the idea of cure by medicine, that drugs actually *cure*, is existent to this day as a remnant of the Middle Ages. A man's death-offense might be that he knew more than he could make others understand about the then secrets of nature. Yet he himself might believe more or less in magic. No one was untouched; all intellect was more or less enslaved.

And when experiments at last began to be made in the mechanisms by which steam might be utilized they were such as boys now make for amusement; such as throwing a steam-jet against the vanes of a paddle-wheel. Such was Branca's engine, made nine years after the landing of our forefathers at Plymouth, and thought worthy of a description and record. The next attempt was much more practical, but cannot be accurately assigned. It consisted of two chambers, from each of which alternately water was forced by steam, and which were filled again by cooling off and the forming of a vacuum where the steam had been. One chamber worked while the other cooled. It was an immense advance in the direction of utility.

About 1698, we begin to encounter the names that are familiar to us in connection with the history of the steam-engine. In that year Thomas Savery obtained a patent for raising water by steam. His was a modification of the idea described above. The boilers used would be of no value now, nevertheless the machine came into considerable use, and the world that learned so gradually became possessed with the idea that there was a utility in the pressure of steam. Savery's engine is said to have grown out of the accident of his throwing a flask containing a little wine on the fire at a tavern. Concluding immediately afterwards that he wanted it, he snatched it off of the fender and plunged it into a basin of water to cool it. The steam inside instantly condensing, the water rushed in and filled it as it cooled.

We now come to the beginning of the steam engine as we understand the term; the machine that involves the use of the cylinder and piston. These two features had been used in pumps long before, the atmospheric pump being one of the oldest of modern machines. The vacuum was known and utilized long before the cause of it was known. [Footnote: The discoverer was an Italian, Torricelli, about 1643. Gallileo, his tutor and friend, did not know why water would not rise in a tube more than thirty-three feet. No one knew of the *weight of the atmosphere*, so late as the early days of this republic. Many did not believe the theory long after that time. Torricelli, by his experiments, demonstrated the fact and invented the mercurial barometer, long known as the "Torricellian Tube." This last instrument led to another discovery; that the weight of the atmosphere varied from time to time in the same locality, and that storms and weather changes were indicated by a rising and falling of the column of mercury in the tube of the siphon-barometer. That which we call the "weather-bureau," organized by General Albert J. Myer, United States Army, in 1870, and growing out of the army signal service, of which he was chief, makes its "forecasts" by the use of the telegraph and the barometer. The "low pressure area" follows a path, which means a change of weather on that path. Notices by telegraph define the route, and the coming storm is not foretold, but *foreknown;* not prophesied, but *ascertained.* If we have been led from the crude pump of Gallileo's time directly to the weather bureau of the present with its invaluable signals to sailors and convenience to everybody, it is no more than is continually to be traced even to the beginning of the wonderful school of modern science.]

But in the beginning it was not proposed to use steam in connection with the cylinder and piston which now really constitutes the steam-engine. Reverting again to the example of the gun, it was suggested to push a piston forward in a tube by the explosion of gunpowder behind it, or to repeat the Savery experiment with powder instead of steam. These ideas were those of about 1678-1685. The very earliest cylinder and piston engine was suggested by Denis Papin in 1690. These early inventors only went a portion of the way, and almost the entire idea of the steam-engine is of much later date. Mankind had then a singular gift of beginning at the wrong end. Every inventor now uses facts that seem to him to have been always known, and that are his by a kind of intuition. But they were all acquired by the tedious experience of a past that is distinguished by a few great names whose owners knew in their time perhaps one-tenth part as much as the modern inventor does, who is unconsciously using the facts learned by old experience. But the others began at the beginning.

[Illustration: EARLY NEWCOMEN PUMPING ENGINE. STEAM-COCK, COLD WATER
COCK AND WASTE-SPIGOT ALL WORKED BY HAND.]

In 1711, almost a hundred years after the arrival at Jamestown and Plymouth of the fathers of our present civilization, the steam-engine that is called Newcomen's began to be used for the pumping of water out of mines. This engine, slightly modified, and especially by the boy who invented the automatic cut-off for the steam valves, was a most rude and clumsy machine measured by our ideas. There appears to have been scarcely a single feature of it that is now visible in a modern engine. The cylinder was always vertical. It had the upper end open, and was a round iron vessel in which a plunger moved up and down. Steam was let in below this plunger, and the walking-beam with which it was connected by a rod had that end of it raised. When raised the steam was cut off, and all that was then under the piston was condensed by a jet of cold water. The outside air-pressure then acted upon it and pushed it down again. In this down-stroke by air-pressure the work was done. The far end of the walking-beam was even counter-weighted to help the steam-pressure. The elastic force of compressed steam was not depended upon, was hardly even known, in this first working and practical engine of the world. Every engine of that time was an experimental structure by itself. The boiler, as we use it, was unknown. Often it was square, stayed and braced against pressure in a most complicated way. Yet the Newcomen engine held its place for about seventy-five years; a very long time in our conception, and in view of the vast possibilities that we now know were before the science. [Footnote: As late as 1880, the steam-engine illustrated and described in the "natural philosophy" text books was still the Newcomen, or Newcomen-Watt engine, and this while that engine was almost unknown in ordinary circumstances, and double-acting high-pressure engines were in operation everywhere. This last, without which not much could be done that is now done, was evidently for a long time after it came into use regarded as a dangerous and un-philosophical experiment, hardly scientific, and not destined to be permanently adopted.]

In the year 1760, James Watt, who was by occupation what is now known as a model-maker, and who lived in Glasgow, was called upon to repair a model of a Newcomen engine belonging to the university. While thus engaged he was impressed with the great waste of steam, or of time and fuel, which is the same thing, involved in the alternate heating and cooling of Newcomen's cylinder. To him occurred the idea of keeping the cylinder as hot as the steam used in it. Watt was therefore the inventor of the first of those economies now regarded as absolute requirements in construction. He made the first "steam-jacket," and was, as well, the author of the idea of covering the cylinder with a coat of wood, or other non-conductor. He contrived a second chamber, outside of the cylinder, where the then indispensable condensation should take place. Then he gave this cylinder for the first time two heads, and let out the piston-rod through a hole in the upper head, with packing. He used steam on the upper side of the piston as well as the lower, and it will be seen that he came very near to making the modern engine.

Yet he did not make it. He was still unable to dispense with the condensing and vacuum and air-pressure ideas. Acting for the first time in the line of real efficiency, he failed to go far enough to attain it. He made a double-acting engine by the addition of many new parts; he even attained the point of applying his idea to the production of circular motion. But he merely doubled the Newcomen idea. His engine became the Newcomen-Watt. He had a condensing chamber at each end of the stroke and could therefore command a reciprocating movement. The walking-beam was retained, not for the purpose for which it is often used now, but because it was indispensable to his semi-atmospheric engine.

[Illustration: THE PERFECTED NEWCOMEN-WATT ENGINE.]

It may seem almost absurd that the universal crank-movement of an engine was ever the subject of a patent. Yet such was the case. A man named Pickard anticipated Watt, and the latter then applied to his engines the "sun-and-planet" movement, instead of the crank, until the patent on the latter expired. The steam-engine marks the beginning of a long series of troubles in the claims of patentees.

In 1782 came Watt's last steam invention, an engine that used steam *expansively*. This was an immense stride. He was also at the same time the inventor of the "throttle," or choke valve, by which he regulated the supply of steam to the piston. It seems a strange thing that up to this time, about 1767, an engine in actual use was started by getting up steam enough to make it go, and waiting for it to begin, and stopped by putting out the fire.

Then he invented the "governor," a contrivance that has scarcely changed in form, and not at all in action, since it was first used, and is one of the few instances of a machine perfect in the beginning. Two balls hang on two rods on each side of an upright shaft, to which the rods are hinged. The shaft is rotated by the engine, and the faster it turns the more the two balls stand out from it. The slower it turns the more they hang down toward it. Any one can illustrate this by whirling in his hands a half-open umbrella. There is a connection between the movement of these balls and the throttle; as they swing out more they close it, as they fall closer to the shaft they open it. The engine will therefore regulate its own speed with reference to the work it has to do from moment to moment.

[Illustration: THE GOVERNOR.]

Through all these changes the original idea remained of a vacuum at the end of every stroke, of indispensable assistance from atmospheric pressure, of a careful use of the direct expansive power of steam, and of the avoidance of the high pressures and the actual power of which steam is now known to be safely capable. [Footnote: In a reputable school "philosophy" printed in 1880, thus: "In some engines" (describing the modern high-pressure engine, universal in most land service) "the apparatus for condensing steam alternately above and below the piston is dispensed with, and the steam, after it has moved the piston from one end of the cylinder to the other, is allowed to escape, by the opening of a valve, directly into the air. To accomplish this it is evident that the steam must have an elastic force greater than the pressure of the air, *or it could not expand and drive out the waste steam on the other side of the piston, in opposition to the pressure of the air.*" According to this teaching, which the young student is expected to understand and to entirely believe, a pressure of steam of, say eighty to a hundred and twenty pounds to the inch on one side of the piston is accompanied by an absolute vacuum there, which permits the pressure of the outside air to exert itself against the opposite side of the piston through the open port at the other end of the cylinder. That is, a state of things which would exist if the steam behind the piston *were suddenly condensed*, exists anyway. If it be true the facts should be more generally known; if not, most of the school "philosophies" need reviewing.] Then an almost unknown American came upon the scene. In English hands the story at once passes from this point to the experiments of Trevethick and George Stevenson with steam as applied to railway locomotion. But as Watt left it and Trevethick found it, the steam engine could never have been applied to locomotion. It was slow, ponderous, complicated and scientific, worked at low pressures, and Watt and his contemporaries would have run away in affright from the innovation that came in between them and the first attempts of the pioneers of the locomotive. This innovation was that of Evans, the American, of whom further presently.

The first steam-engine ever built in the United States was probably of the Watt pattern, in 1773. In 1776, the year of beginning for ourselves, there were only two engines of any kind in the colonies; one at Passaic, N. J., the other at Philadelphia. We were full of the idea of the independence we had won soon afterwards, but in material respects we had all before us.

In 1787, Oliver Evans introduced improvements in grain mills, and was generally efficient as one of the beginners in the field of American invention. Soon afterwards he is known to have made a steam-engine which was the first high-pressure double-acting engine ever made. The engine that used steam at each end of the cylinder with a vacuum and a condenser, was in this first instance, so far as any record can be found, supplanted by the engine of to-day. The reason of the delay it is difficult to account for on any other grounds than lack of boldness, for unquestionably the early experimenters knew that such an engine could be made. They were afraid of the power they had evoked. Such a machine may have seemed to them a willful toying with disaster. Their efforts were bent during many years toward rendering a treacherous giant useful, yet entirely harmless. Their boilers, greatly improved over those I have mentioned, never were such as were afterwards made to suit the high pressures required by the audacity of Hopkins. This audacity was the mother of the locomotive, and of that engine which almost from that date has been used for nearly every purpose of our modern life that requires power. The American innovation may have passed unnoticed at the time, but intentionally or otherwise it was imitated as a preliminary to all modern engines. Nearly a century passed between the making of the first practical engine and that one which now stands as the type of many thousands. But now every little saw-mill in the American woods could have, and finally did have, its little cheap, unscientific, powerful and non-vacuum engine, set up and worked without experience, and maintained in working order by an unskilled laborer. A thousand uses for steam grew out of this experiment of a Yankee who knew no better than to tempt fate with a high-pressure and speed and recklessness that has now become almost universal.

There was with Watt and his contemporaries apparently a fondness for cost and complications. Most likely the finished Watt engine was a handsome and stately machine, imposing in its deliberate movements. There is apparently nothing simpler than the placing of the head of the piston-rod between two guide-pieces to keep it in line and give it bearing. Yet we have only to turn back a few years and see the elaborate and beautiful geometrical diagram contrived by Watt to produce the same simple effect, and known as a "parallel motion." It kept its place until the walking-beam was cast away, and the American horizontal engine came into almost universal use.

The object of this chapter so far has been to present an idea of beginnings; of the evolution of the universal and indispensable machine of civilization. The steam-engine has given a new impetus to industry, and in a sense an added meaning to life. It has made possible most that was ever dreamed of material greatness. It has altered the destiny of this nation, and other nations, made greatness out of crude beginnings, wealth out of poverty, prosperity upon thousands of square miles of uninhabitable wilderness. It was the chiefest instrumentality in the widening of civilization, the bringing together of alien peoples, the dissemination of ideas. Electricity may carry the idea; steam carries the man with the idea. The crude misconceptions of old times existed naturally before its time, and have largely vanished since it came. Marco Polo and Mandeville and their kind are no longer possibilities. Applied to transportation, locomotion alone, its effects have been revolutionary. Applied to common life in its minute ramifications these effects could not have been believed or foretold, and are incredible. The thought might be followed indefinitely, and it is almost impossible to compare the world as we know it with the world of our immediate ancestors. Only by means of contrasts, startling in their details, can we arrive at an adequate estimate, even as a moral farce, of the power of steam as embodied in the modern engine in a thousand forms.

* * * * *

Perhaps it might be well to attempt to convey, for the benefit of the youngest reader, an idea of the actual working of the machine we call a steam-engine. There are hundreds of forms, and yet they are all alike in essentials. To know the principle of one is to know that of all. There is probably not an engine in the world in effective common use—the odd and unusual rotary and other forms never having been practical engines—that is not constructed upon the plan of the cylinder and piston. These two parts make the engine. If they are understood only differences in construction and detail remain.

Imagine a short tube into which you have inserted a pellet, or wad of any kind, so that it fits tolerably, yet moves easily back and forth in the bore of the tube. If this pellet or wad is at one end of the tube you may, by inserting that end in your mouth and putting air-pressure upon it, make it slide to the other end. You do not touch it with anything; you may push it back and forth with your breath as many times as you wish, not by blowing against it, so to speak, but by producing an actual air-pressure upon it which is confined by the sides of the tube and cannot go elsewhere. The only pressure necessary is enough to move the pellet.

Now, if you push this little pellet one way by the air-pressure from your mouth, and then, instead of reversing the tube in the mouth and pushing it back again in the same way, reverse the process and suck the air out from behind it, it comes back by the pressure of the outside atmosphere. This was the way the first steam engines worked. Their only purpose was to get the piston lifted, and air-pressure did all the actual work.

If you turn the tube, and put an air-pressure first at one end and then at the other, and pay no attention to vacuum or atmospheric pressure, you will have the principle of the later modern, almost universal, high-pressure, double-acting steam-engine.

But now you must imagine that the tube is fixed immovably, and that the air-pressure is constant in a pipe leading to the tube, and yet must be admitted first to one end of the tube and then to the other alternately, in order to push the pellet back and forth in it. It seems simple. Perhaps the young reader can find a way to do it, but it required about a hundred years for ingenious men to find out how to do precisely the same thing automatically. It involves the steam-chest and the slide-valve, and all other kinds of steam valves that have been invented, including the Corliss cut-off, and all others that are akin to it in object and action.

But now imagine the tube closed at each end to begin with, and the little moving pellet, or plunger, on the inside. To get the air into both ends of the tube alternately, and to use its pressure on each side of the pellet, we will suppose that the air-pipe is forked, and that one end of each fork is inserted into the side of the tube near the end, like the figure below, and imagine also that you have put a finger over each end of the tube.

[Illustration: Fig. 1]

We are now getting the air-pressure through the pipe in both ends of the tube alike, and do not move the pellet either way. To make it move we must do something more, and open one end of the tube, and close that fork of the air-pipe, and thus get all the pressure on one side of the pellet. Remove one finger from the end of the tube, and pinch the fork of the air-tube that is on that side. The pellet will now move toward that end of the tube which is open. Reverse the process, and it can be pushed back again with air-pressure to the other end, and so on indefinitely.

Let us improve the process. We will close each end of the tube permanently, and insert four cocks in the tube and forked pipe.

We have here two tubes inserted at each end of the large tube, and in each of these is a cock. We have each cock connected by a rod to the lever set on a pin in the middle of the tube. We must have these cocks so arranged that when the lever is moved (say) to the right, A. is opened and B. is closed, and D. is opened and C. is closed. Now if the air-pressure is constant through the forked air-tube, and the cock E. is open, if the top of the lever is moved to the right, the pellet will be pushed to the left in the large tube. If the lever is moved to the left, and the two cocks that were open are closed, and the two that were closed are opened again, the pellet will be sent back to the other end of the tube. This movement of the pellet in the tube will occur as often as the lever is moved and there is any air-pressure in the forked tube. There is a *supply*-cock, opened and an *escape*-cock closed, and an escape-cock *opened* and a supply-cock *closed*, at each end of the tube, *every time the lever is moved*.

[Illustration: Fig. 2]

We are using air instead of steam, and the movement of these four cocks all at the same time, and the result of moving them, is precisely that of the slide-valve of a steam-engine. The diagrams of this slide-valve would be difficult to understand. The action of the cocks can be more readily understood, and the result, and even much of the action, is precisely the same.

But to make the arrangement entirely efficient we must go a little further into the construction of a steam-engine. The pellet in the tube has no connection with the outside, and we can get nothing from it. So we give it a stem, thus: and when we do so we change it into a piston and its rod. Where it passes through the stopper at the end of the tube it must pass air- (or steam-) tight. Then as we push the piston back and forth we have a movement that we can attach to machinery at the end of the rod, and get a result from. We also move the cocks, or valves, automatically by the movement of the rod.

[Illustration: Fig. 3]

Turning now to Fig. 3 again let us imagine a connection made between the rod and the end of the lever in Fig. 2. Now put on the air (or steam) pressure, and when the piston has reached the right-hand end of the tube it automatically, by its connections, closes B. and opens A., and opens D. and closes C. The pellet will be pushed back in the tube and go to the other end of it, through the pressure coming against the piston through the part of the air tube where the cock D. is open. It reaches the left-hand end of the tube, and we must imagine that when it gets there it, in the same manner and by the proper connections, closes D., opens C., closes A. and opens B. If these mechanical movements are completed it must be plain that so long as the air (or steam) pressure is continued in the forked pipe the piston will automatically cut off its supply and open its escape at each alternate end, and move back and forth. Any boy can see how a backward and forward movement may be made to give motion to a crank. All other details in an engine are questions of convenience in construction, and not questions of principle or manner of action.

Of older readers, I might request the supposition that, in Fig. 2, only the valves A. and B. were automatically and invariably opened and closed by the action of the piston-rod of Fig. 3, and that C. and D. were controlled solely by the governor, before mentioned, which we will suppose to be located at E. Then the escape of the steam ahead of the piston must always come at the same time with reference to the stroke, but the supply will depend upon the requirements of each individual stroke, and the work it has to do, and afford to the piston a greater or less push, as the emergencies of that particular instant may require. This arrangement would be one of regularity of movement and of economy in the use of steam. That which is needed is supplied, and no more. This is the principle and the object of the Corliss cut-off, and of all others similar to it in purpose. Their principle is that *only the escape is automatically controlled by the movements of the piston-rod*, occurring always at the same time with reference to the stroke, while *the supply is under control of the movement of the governor*, and regulated according to the emergencies of the movement. The governor, in any of its forms, as ordinarily applied, performs only half of this function. It regulates the general supply of steam to the cylinder, but the supply-valve continues to be opened, always to full width, and always at the same moment with reference to the stroke. With the two separate sets of automatic machinery required by engines of the Corliss type, the piston does not always receive its steam at the beginning of the stroke, and the supply may be cut off partially or entirely at any point in its passage along the cylinder, as the work to be done requires. The economic value of such an arrangement is manifest. No attempt is made here to explain by means of elaborate diagrams. It is believed that if the reason of things, and the principle of action, is clear, the particulars may be easily studied by any reader who is disposed to master mechanical details.

THE AGE OF STEEL

In very recent times the processes of civilization have had a strong and almost unnoted tendency toward the increased use of the *best*. Thus, most that iron once was, in use and practice, steel now is. This use, growing daily, widens the scope that must be taken in discussing the features of an Age of Steel. One name has largely supplanted the other. In effect iron has become steel. Had this chapter been written twenty, or perhaps ten, years earlier, it should have been more appropriately entitled the Age of Iron. A separation of the two great metals in general description would be merely technical, and I shall treat the subject very much as though, in accordance with the practical facts of the case, the two metals constituted one general subject, one of them gradually supplanting the other in most of the fields of industry where iron only was formerly used.

The greatest progresses of the race are almost always unappreciated at the time, and are certainly undervalued, except by contrast and comparison. We must continually turn backward to see how far we have gone. An individual who is born into a certain condition thinks it as hard as any other until by experience and comparison he discovers what his times might have been. As for us, in the year 1894, we are not compelled to look backward very far to observe a striking contrast.

[Illustration: IN OLD TIMES. PRYING OUT A "BLOOM."]

All the wealth of today is built upon the forests and prairies and swamps of yesterday, and we must take a wider and more comprehensive glance backward if we should wish to institute those comparisons which make contrasts startling.

We are accustomed to read and to hear of the "Age" of this or that. There was a "Stone" Age, beginning with the tribes to whom it came before the beginnings of their history, or even of tradition, and if we look far backward we may contrast our own time with the times of men who knew no metals. They were men. They lived and hoped and died as we do, even in what is now our own country. Often they were not even barbarians. They built houses and forts, and dug drains and built aqueducts, and tilled the soil. They knew the value of those things we most value now, home and country; and they organized armies, and fought battles, and died for an idea, as we do. Yet all the time, a time ages long, the utmost help they had found for the bare and unaided hand was the serrated edge of a splintered flint, or the chance-found fragment beside a stream that nature, in a thousand or a million years of polishing, had shaped into the rude semblance of a hammer or a pestle. All men have in their time burned and scraped and fashioned all they needed with an astonishing faculty of making it answer their needs. They once almost occupied the world. Such were those who, so far as we know, were once the exclusive owners of this continent. They were an agricultural, industrious and home-loving people. [Footnote: The Mound Builders and Cave Dwellers. They knew only lead and copper.]

Then came, with a strange leaving out of the plentiful and easily worked metals which are the subject of this chapter, the great Age of Bronze. This next stage of progress after stone was marked by a skillful alloy, requiring even now some scientific knowledge in its compounding of copper and tin. A thousand theories have been brought forward to account for this hiatus in the natural stages of human progress, the truth probably being that both tin and copper are more fusible than iron-ores, and that both are found as natural metals. Some accident such as accounts for the first glass, [Footnote: The story is told by Pliny. Some sailors, landing on the eastern coast of Spain, supported their cooking utensils on the sand with stones, and built a fire under them. When they had finished their meal, glass was found to have been made from the niter and sea-sand by the heat of their fire. The same thing has been done, by accident, in more recent times, and may have been done before the incident recounted. It is also done by the lightning striking into sand and making those peculiar glass tubes known as *Fulmenites*, found in museums and not very uncommon.] some camp-fire unintended fusion, produced the alloy that became the metal of all the arms and arts, and so remained for uncounted centuries. In this connection it is declared that the Age of Bronze knew something that we cannot discover; the art of tempering the alloy so that it would bear an edge like fine steel. If this be true and we could do it, we should by choice supplant the subject of this chapter for a thousand uses. As the matter stands, and in our ignorance of a supposed ancient secret, the tempering of bronze has an effect precisely opposite to that which the process has upon steel.

Nevertheless, the old Age of Bronze had its vicissitudes. Those men knew nothing that we consider knowledge now. It was a time when some of the most splendid temples, palaces and pyramids were constructed, and these now lie ruined yet indestructible in the nooks and corners of a desert world. Perhaps the hard rock was chiselled with tools of tempered copper. The fact is of little importance now since the object of the art is almost unknown, and the scattered capitals and columns of Baalbeck are like monuments without inscriptions; the commemorating memorials of a memory unknown. The Age of Bronze and all other ages that have preceded ours lacked the great essentials that insure perpetuity. The Age of Steel, that came last, that is ours now; a degenerate time by all ancient standards; has for its crowning triumph a single machine which is alone enough to satisfy the union of two names that are to us what Caster and Pollux were to the bronze-armed Roman legions of the heroic time—the modern power printing-press.

It may be well to ask and answer the question that at the first view may seem to the reader almost absurd. What is steel? The answer must, in the majority of instances, be given in accordance with the common conception; which is that it is not iron, yet very like it. The old classification of the metal, even familiarly known, needs now to be supplemented, since it does not describe the modern cast and malleable compounds of iron, carbon and metalloids used for structural purposes, and constituting at least three-fourths of the metal now made under the name of steel. The old term, steel, meant the cast, but malleable, product of iron, containing as much carbon as would cause the metal to harden when heated to redness and quenched in water. It must also be included in the definition that the product must be as free as possible from all admixtures except the requisite amount of carbon. This is "tool" steel. [Footnote: It must not be understood that tool steel was always a cast metal. In manufacturing, iron bars were laid together in a box or retort, together with powdered charcoal, and heated to a certain degree for a certain time. The carbon from the charcoal was absorbed by the iron, and from the blistered appearance of the bars when taken out this product was, and is known as "blister" steel.]

And here occurs a strange thing. A skill in chemistry, the successor of alchemy, is the educational product of the highest form of civilization.

[Illustration: ANCIENT SMELTING. A RUDE WALL ENCLOSING ALTERNATE LAYERS
OF IRON ORE AND CHARCOAL.]

Metallurgy is the highest and most difficult branch of chemistry. Steel is the best result of metallurgy. Yet steel is one of the oldest products of the race, and in lands that have been asleep since written history began. Wendell Phillips in a lecture upon "The Lost Arts,"— celebrated at the date of its delivery, but now obsolete because not touching upon advances made in science since Phillips's day,—states that the first needle ever made in England, in the time of Henry VIII, was made by a Negro, and that when he died the art died with him. They did not know how to prepare the steel or how to make the needle. He adds that some of the earliest travelers in Africa found a tribe in the interior who gave them better razors than the explorers had. Oriental steel has been celebrated for ages as an inimitable product. It is certainly true that by the simple processes of semi-barbarism the finest tool-steel has been manufactured, perhaps from the days of Tubal Cain downward. The keenness of edge, the temper whose secret is now unknown, the marvelous elasticity of the tools of ancient Damascus, are familiar by repute to every reader and have been celebrated for thousands of years. The swords and daggers made in central Asia two thousand years ago were more remarkable than any similar product of the present for elaborate and beautiful finish as well as for a cutting quality and a tenacity of edge unknown to modern days. All the tests and experiments of a modern government arsenal, with all the technical knowledge of modern times, do not produce such tool-steel. It is also alleged that the ancient weapons did not rust as ours do, and that the oldest are bright to this day. The steel tools and arms that are made in the strange country of India do not rust there, while in the same climate ours are eaten away. Besides the secret of tempering bronze, it would seem that among the lost arts [Footnote: Modern science dates from three discoveries. That of Copernicus, the effect of which was to separate scientific astronomy, the astronomy of natural law and defined cause, from astrology, or the astronomy of assertion and tradition. That of Torricelli and Paschal of the actual and measurable weight of the atmosphere, which was the beginning for us of the science of physics, and that of Lavoisier who suspected, and Priestly who demonstrated, oxygen and destroyed the last vestiges of the theory of alchemy. Stahl was the last of these, and Lavoisier the first of the new school in that which I have stated is the highest development of modern science, chemistry. In all these departments we have no adequate reason to assert that we are not ourselves mere students. Some of the functions of oxygen, and the simplest, were unknown within five years before the date of these chapters.]—a subject that it is easy to make too much of—there was a chemical ingredient or proportion in steel that we now know nothing of. The old lands of sameness and slumber have kept their secrets.

The definition of the word "steel" has been the subject of a scientific quarrel on account of new processes. The grand distinguishing trait of steel, to which it owes all the qualities that make it valuable for the uses to which no other metal can be put, is *homogeneity due to fusion*. Wrought iron, while having similar chemical qualities, and often as much carbon, is *laminated in structure*. Structural qualities are largely increasing in importance, and as the structural compounds came gradually to be produced more and more by the casting processes; as they ceased to be laminated in structure and became homogeneous, they were called by the name of steel. The name has been based upon the structure of the material rather than upon its chemical ingredients as heretofore. There is now a disposition to call all compounds of iron that are crystalline in structure, made homogeneous by casting, by the general name of steel, and to distinguish all those whose structural quality is due to welding by the name of iron. [Footnote: It should be understood that the shapes of structural and other forms of what we now call steel are given by rolling the ingot after casting, and that the crystalline composition of the metal remains.] This is an outline of the controversy about the differences which should be expressed by a name, between tool steel and structural steel. In tool steel there is an almost infinite variety as to quality. The best is a high product of practical science, and how to make the best seems now, as hinted above, a lost art. It has, besides, a great variety. These varieties are only produced after thousands of experiments directed to finding out what ingredients and processes make toward the desired result. These processes, were they all known outside the manufactories of certain specialists, would little interest the general reader. All machinists know of certain brands of tool steel which they prefer. Tool steel is made especially for certain purposes; as for razors and surgical instruments, for saws, for files, for springs, for cutting tools generally. In these there may be little actual difference of quality or manufacture. The tempering of steel after it has been forged into shape is a specialty, almost a natural gift. The manufacture of tool steel, is, as stated, one of the most technical of the arts, and one of the most complicated of the applications of long experience and experiment.

Cast steel was first made in 1770 by Huntsman, who for the first time melted the "blistered" steel, which until that time had been the tool steel of commerce, in a crucible. Since that time the process of melting wrought iron has become practical and cheap, and results in *crystalline*, instead of a laminated structure for all steels. The definition of steel now is that it is *a compound of iron which has been cast from a fluid state into a malleable mass.*

The ordinary test applied to distinguish wrought iron from steel is to ascertain whether the metal hardens with heating and suddenly cooling in cold water, becoming again softened on reheating and cooling slowly. If it does this it is steel of some quality, good or bad; if not, it is iron.

* * * * *

The first mention of iron-ore in America is by Thomas Harriot, an English writer of the time of Raleigh's first colonies. He wrote a history of the settlement on Roanoke Island, in which he says: "In two places in the countrey specially, one about foure score and the other six score miles from the port or place where wee dwelt, wee founde neere the water side the ground to be rockie, which by the triall of a minerall man, was found to hold iron richly. It is founde in manie places in the countrey else." Harriot speaks further of "the small charge for the labour and feeding of men; the infinite store of wood; the want of wood and the deerness thereof in England." It was before the day of coal and coke, or of any of the processes known now. The iron mines of Roanoke Island were never heard of again.

Iron-ore in the colonies is again heard of in the history of Jamestown, in 1607. A ship sailed from there in 1608 freighted with "iron-ore, sassafras, cedar posts and walnut boards." Seventeen tons of iron were made from this ore, and sold for four pounds per ton. This was the first iron ever made from American ores. The first ironworks ever erected in this country were, of course almost, burned by the Indians, in 1622, and in connection three hundred persons were killed.

[Illustration: EARLY SMELTING IN AMERICA.]

Fire and blood was the end of the beginning of many American industries. Ore was plentiful, wood was superabundant, methods were crude. They could easily excel the Virginia colonists in making iron in Persia and India at the same date. The orientals had certain processes, descended to them from remote times, discovered and practiced by the first metal-workers that ever lived. The difference in the situation now is that here the situation and methods have so changed that the story is almost incredible. There, they remain as always. The first instance of iron-smelting in America is a text from which might be taken the entire vast sermon of modern industrial civilization.

The orientals lacked the steam-engine. So did we in America. The blast was impossible everywhere except by hand, and contrivances for this purpose are of very great antiquity. The bellows was used in Egypt three thousand years ago. It may be that the very first thought by primitive man was of how to smelt the metals he wanted so much and needed so badly. His efforts to procure a means of making his fire burn under his little dump of ore led him first into the science which has attained a new importance in very recent times, pneumatics. The first American furnaces were blown by the ordinary leather bellows, or by a contrivance they had which was called a "blowing tub," or by a very ancient machine known as a *"trompe"* in which water running through a wooden pipe was very ingeniously made to furnish air to a furnace. It is when the means are small that ingenuity is actually shown. If the later man is deprived of the use of the latest machinery he will decline to undertake an enterprise where it is required. The same man in the woods, with absolute necessity for his companion, will show an astonishing capacity for persevering invention, and will live, and succeed.

[Illustration: WATER-POWER BLOWING TUB.]

In the lack of steam they learned, as stated, to use water-power for making the blast. The "blowing-tub" was such a contrivance. It was built of wood, and the air-boxes were square. There were two of these, with square pistons and a walking-beam between them. A third box held the air under a weighted piston and fed it to the furnace. Some of these were still in effective use as late as 1873. They were still used long after steam came. The entire machine might be called, correctly, a very large piston-bellows. A smaller machine with a single barrel may be found now, reduced, in the hands of men who clean the interior of pianos, and tune them.

The first iron works built in the present United States that were commercially successful, were established in Massachusetts, in the town of Saugus, a few miles from Boston. The company had a monopoly of manufacture under grant for ten years. [Footnote: Some quaint records exist of the incidents of manufacturing in those times.

In 1728, Samuel Higley and Joseph Dewey, of Connecticut, represented to the Legislature that Higley had, "with great pains and cost, found out and obtained a curious art by which to convert, change, or transmute, common iron into good steel sufficient for any use, and was the first that ever performed such an operation in America." A certificate, signed by Timothy Phelps and John Drake, blacksmiths, states that, in June, 1725, Mr. Higley obtained from the subscribers several pieces of iron, so shaped that they could be known again, and that a few days later "he brought the same pieces which we let him have, and we proved them and found them good steel, which was the first steel that ever was made in this country, that we ever saw or heard of." But this remarkable transmuting process was not heard of again unless it be the process of "case-hardening," re-invented some years ago, and known now to mechanics as a recipe.

The smallness of things may be inferred from the fact that, in 1740, the Connecticut Legislature granted to Messrs. Fitch, Walker & Wyllys "the sole privilege of making steel for the term of fifteen years, upon this condition that they should, in the space of two years, make half a ton of steel." Even this condition was not complied with and the term was extended.] They began in 1643, twenty-three years after the landing, which is one of the evidences of the anxiety of those troublesome people to be independent, and of how well men knew, even in those early times, how much the production of iron at home has to do with that independence. This new industry was, at all times, controlled and regulated by law.

The very first hollow-ware casting made in America is said to be still in existence. It was a little kettle holding less than a quart.

[Illustration: THE FIRST CASTING MADE IN AMERICA.]

The beginnings of the iron industry in America were none too early. There came a need for them very soon after they had extended into other parts of New England, and into New Jersey, New York, Pennsylvania and Maryland. In 1775, there were a large number of small furnaces and foundries. But coal and iron, the two earth-born servants of national progress which are now always twins, were not then coupled. The first of them was out of consideration. The early iron men looked for water-falls instead, and for the wood of the primeval forest. [Footnote: It is now easy to learn that a coal-mine may be a more valuable possession than a gold-mine, and that iron is better as an industry than silver. There are mountains of iron in Mexico, but no coal, and silver-mines so rich that silver, smelted with expensive wood fuel, is the staple product of the country. Yet the people are among the poorest in Christendom. There is a ceaseless iron-famine, so that the chiefest form of railway robbery is the stealing of the links and pins from trains. There are almost no metal industries. A barbaric agriculture prevails for the want of material for the making of tools. The actual means of progress are not at hand, notwithstanding the product of silver, which goes by weight as a commodity to purchase most that the country needs.] They became very necessary to the country in 1755—when the "French" war came, and they then began the making of the shot and guns used in that struggle, and became accustomed to the manufacture in time for the Revolution. Looking back for causes conducive to momentous results, we may here find one not usually considered in the histories. But for the advancement of the iron industry in America, great for the time and circumstances, independence could not have been won, and even the *feeling* and desire of independence would have been indefinitely delayed.

The industry was slow, painful, and uncertain, only because the mechanic arts were pursued only to an extent possible with the skill and muscular energy of men. There were none of the wonderful automatic mechanisms that we know as machine-tools. There was only the almost unaided human arm with which to subdue the boundless savagery of a continent, and win independence and form a nation besides. The demand for huge masses of the most essential of the factors of civilization has grown since, because the ironclad and the big gun have come, and those inadequate forces and crude methods supplied for a time the demand that was small and imperative. The largest mass made then, and frequently spoken of in colonial records, was a piece called a "sow;" spelled then "sowe." It was a long, triangular mass, cast by being run into a trench made in sand. [Footnote: When, later, little side-trenches were made beside the first, with little channels to carry the metal into them, the smaller castings were naturally called "pigges." Hence our "pig-iron."]

[Illustration: MAKING A TRENCH TO CAST A "SOWE."]

Those were the palmy days of the "trip hammer." Nasmyth was not born until 1808, and no machine inventor had yet come upon the scene. The steam-hammer that bears his name, which means a ponderous and powerful machine in which the hammer is lifted by the direct action of steam in a piston, the lower end of whose rod is the hammer-head, has done more for the development of the iron industry than any other mechanical invention. It was not actually used until 1842, or '43. It finally, with many improvements in detail, grew into a monster, the hammer-head, or "tup," being a mass of many tons. And they of modern times were not content merely to let this great mass fall. They let in steam above the piston, and jammed it down upon the mass of glowing metal, with a shock that jars the earth. The strange thing about this Titanic machine is that it can crack an egg, or flatten out a ton or more of glowing iron. Hundreds of the forgings of later times, such as the wrought iron or steel frames of locomotives, and the shafts of steamers, and the forged modern guns, could not be made by forging without this steam hammer.

[Illustration: THE STEAM HAMMER.]

Then slowly came the period of all kinds of "machine tools." During the period briefly described above they could not make sheet metal. The rolling mill must have come, not only before the modern steam-boiler, but even before the modern plow could be made. Can the reader imagine a time in the United States when sheet metal could not be rolled, and even tin plates were not known? If so, he can instantly transport himself to the times of the wooden "trencher," and the "pewter" mug and pitcher, to the days when iron rails for tramways were unknown, and when even the "strap-iron," always necessary, was rudely and slowly hammered out on an anvil. [Footnote: About 1720, nails were the most needed of all the articles of a new country. Farmers made them for themselves, at home. The secret of how to roll out a sheet and split it into nail-rods was stolen from the one shop that knew how, at Milton, Mass., to give to another at Mlddleboro. The thief had the Biblical name of Hashay H. Thomas. He stole the secret while the hands of the Milton mill were gone to dinner, and served his country and broke up a small monopoly in so doing.]

Shears came with the "rolls;" vast engines of gigantic biting capacity, that cut sheets of iron as a lady's scissors cut paper. This cut the squares of metal used for boiler plates, and the steam-engine having come, was turned to the manufacture of materials for its own construction. Others were able to bite off great bars.

The first mill in which iron was rolled in America, was built in 1817 near Connellsville, in Fayette county, Penn. Until 1844, the rolling mills of this country produced little more than bar-iron, hoops, and plates. All the early attempts at railroads used the "strap" rail; unless cast "fish-bellies" were used; which was flat bar-iron provided with counter sunk holes, in which to drive nails for holding the iron to long stringers of wood laid upon ties. When actual rail-making for railroads began, the rolling mill raised its powers to meet the emergency. The "T" rail, universally now used, was invented by Robert Stevens, president and chief engineer of the Camden and Amboy railroad, and the first of them were laid as track for that road in 1832. From this time until 1850, rolling mills for making "U" and "T" rails rapidly increased in number, but in that year all but two had ceased to be operated because of foreign competition.

[Illustration: SHEARS FOR CUTTING BAR-IRON.]

During some five years previous to this writing a revolution has taken place in the construction of buildings which has resulted in what is known as the "sky-scraper." This was, in many respects, the most startling innovation of times that are startling in most other respects, and was begun in that metropolis of surprises and successes, the city of Chicago. This innovation was really such in the matter of using steel in the entire framing of a commercial building, but it was not the first use of metal as a building material. The first iron beams used in buildings were made in 1854, in a rolling mill at Trenton, N. J., and were used in the construction of the Cooper Institute, and the building of Harper & Brothers. For these special rolls, of a special invention, were made. These have now become obsolete, and a new arrangement is used for what are known as "structural shapes."

[Illustration: HYDRAULIC SHEARS. THE KNIFE HAS A PRESSURE OF 3,000 TONS,
CLIPPING PIECES OF IRON TWO BY FOUR FEET.]

I have spoken of the use of wood-fuel in the early stages of iron manufacture in this country, followed by the adoption exclusively of coal and its products. Then, many years later, came the departure from this in the use of gas for fuel. The first use of this kind is said to date as far back as the eighth century, and modifications of the idea had been put in practice in this country, in which gas was first made from coal and then used as fuel. Then came "natural gas." This product has been known for many centuries. It was the "eternal" fuel of the Persian fire-worshippers, and has been used as fuel in China for ages. Its earliest use in this country was in 1827, when it was made to light the village of Fredonia, N. Y. Probably its first use for manufacturing purposes was by a man named Tompkins, who used it to heat salt-kettles in the Kenawha valley in 1842. Its next use for manufacturing purposes was made in a rolling mill in Armstrong county, Penn., in 1874, forty-seven years after it had been used at Fredonia, and twenty-nine years after it had been used to boil salt.

Now the use of natural gas as manufacturing fuel is universal, not alone over the spot where the gas is found, but in localities hundreds of miles away. It is one of the strangest developments of modern scientific ingenuity. That enormous battery of boilers, which was one of the most imposing spectacles of the Columbian Exhibition of 1893, whose roar was like that of Niagara, was fed by invisible fuel that came silently in pipes from a state outside of that where the great fair was held. We are left to the conclusion that the making of the coal into gas at the mine, and the shipping of it to the place of consumption through pipes, is more certain of realization than were a hundred of the early problems of American progress that have now been successful for so long that the date of their beginning is almost forgotten.

THE STEEL OF THE PRESENT.—The story of steel has now almost been told, in that general outline which is all that is possible without an extensive detail not interesting to the general reader. In it is included, of necessity, a resumé of the progress, from the earliest times in this country, of the great industry which is more indicative than any other of the material growth of a nation. I now come to that time when steel began to take the place that iron had always held in structural work of every class. The differences between this structural steel and that which men have known by the name exclusively from remote ages, I have so far indicated only by reference to the well-known qualities of the latter. It now remains to describe the first.

In 1846 an American named William Kelley was the owner of an iron-works at Eddyville, Ky. It was an early era in American manufactures of all kinds, and the district was isolated, the town not having five hundred inhabitants, and the best mechanical appliances were remote.

In 1847, Kelley began, without suggestion or knowledge of any experiments going on elsewhere, to experiment in the processes now known as the "Bessemer," for the converting of iron into steel. To him occurred, as it now appears first, the idea that in the refining process fuel would be unnecessary after the iron was melted if *powerful blasts of air were forced into the fluid metal*. This is the basic principle of the Bessemer process. The theory was that the heat generated by the union of the oxygen of the air with the carbon of the metal, would accomplish the refining. Kelley was trying to produce malleable iron in a new, rapid and effective way. It was merely an economy in manufacture he was endeavoring to attain.

To this end he made a furnace into which passed an air-blast pipe, through which a stream of air was forced into the mass of melted metal. He produced refined iron. Following this he made what is now called a "converter," in which he could refine fifteen hundred pounds of metal in five minutes, effecting a great saving in time and fuel, and in his little establishment the old processes were thenceforth dispensed with. It was locally known as "Kelley's air-boiling process." It proved finally to be the most important, in large results, ever conceived in metallurgy. I refer to it hurriedly, and do not attempt to follow the inventor's own description of his constructions and experiments. When he heard that others in England were following the same line of experiment, he applied for a patent. He was decided to be the first inventor of the process, and a patent was granted him over Bessemer, who was a few days before him. There is no question that others were more skillful, and with better opportunities and scientific associations, in carrying out the final details, mechanical and chemical, which have completed the Kelley process for present commercial uses. Neither is there any question that this back-woods iron-making American was the first to refine iron by passing through it, while fluid, a stream of air, which is the process of making that steel which is not tool steel, and yet is steel, the now almost universal material for the making of structures; the material of the Ferris wheel, the wonderful palaces of the Columbian exposition, the sky-scrapers of Chicago, the rails, the tacks, [Footnote: In the history of Rhode Island, by Arnold, it is claimed that the first cold cut nails in the world were made by Jeremiah Wilkinson, in 1777. The process was to cut them from an old chest-lock with a pair of shears, and head them in a smith's vise. Then small nails were cut from old Spanish hoops, and headed in a vise by hand. Needles and pins were made by the same person from wire drawn by himself. Supposing this to be the beginning of the cut-nail idea, *the machine for making them* would still remain the actual and practical invention, since it would mark the beginning of the industry as such. The importance of the latter event may be measured by the fact that about the end of the last century there began a strong demand. In the homely farm-houses, or the little contracted shops of New England villages, the descendants of the Pilgrims toiled providently, through the long winter months, at beating into shape the little nails which play so useful a part in modern industry. A small anvil served to beat the wire or strip of iron into shape and point it; a vise worked by the foot clutched it between jaws furnished with a gauge to regulate the length, leaving a certain portion projecting, which, when beaten flat by a hammer, formed the head. This was industry, but not manufacture, for in 1890 the manufacturers of this country produced over *eight hundred million pounds* of iron, steel, and wire nails, representing a consumption of this absolutely indispensable manufacture for that year, at the rate of over *twelve pounds* for each individual inhabitant of the United States.] the fence-wire, the sheet-metal, the rails of the steam-railroads and the street-lines, the thousand things that cannot be thought of without a list, and which is a material that is furnished more cheaply than the old iron articles were for the same purposes.

[Illustration: SECTIONAL VIEW OF A BESSEMER "CONVERTER."]

The technical detail of steel-making is exceedingly interesting to students of applied science, but it *is* detail, the key to which is in the process mentioned; the forcing of a stream of air through a molten mass of iron. The "converter" is a huge pitcher-shaped vessel, hung upon trunnions so as to be tilted, and it is usual to admit through these trunnions, by means of a continuing pipe, the stream of air. The converters may contain ten tons or more of liquid metal at one time, which mass is converted from iron into steel at one operation.

Forty-five years ago, or less, works that could turn out fifty tons of iron in a day were very large. Now there are many that make *five hundred tons* of steel in the same time. Then, nearly all the work was done by hand, and men in large numbers handled the details of all processes. Now it would be impossible for human hands and strength to do the work. The steel-mill is, indeed, the most colossal combination of Steam and Steel. There are tireless arms, moved by steam, insensible alike to monstrous strains and white heat, which seize the vast ingots and carry them to and fro, handling with incredible celerity the masses that were unknown to man before the invention of the Bessemer process. And all these operations are directed and controlled by a man who stands in one place, strangely yet not inappropriately named a "pulpit," by means of the hand-gear that gives them all to him like toys.

No one who has seen a steel-mill in operation, can go away and really write a description of it; no artist or camera has ever made its portrait, yet it is the most impressive scene of the modern, the industrial, world. There is a "fervent heat," surpassing in its impressions all the descriptions of the Bible, and which destroys all doubt of fire with capacity to burn a world and "roll the heavens together as a scroll." There is a clang and clatter accompanying a marvelous order. There are clouds of steam. There are displays of sparks and glow surpassing all the pyrotechnics of art. Monstrous throats gasp for a draught of white-hot metal and take it at a gulp. Glowing masses are trundled to and fro. There are mountains of ore, disappearing in a night, and ever renewed. There is a railway system, and the huge masses are conveyed from place to place by locomotive engines. There is a water system that would supply a town. There may be miles of underground pipes bringing gas for fuel. Amid these scenes flit strong men, naked to the waist, unharmed in the red pandemonium, guiding every process, superintending every result; like other men, yet leading a life so strange that it is apparently impossible. The glowing rivers they escape; corruscating showers of flying white-hot metal do not fall upon them; the leaping, roaring, hungry, annihilating flames do not touch them; the gurgling streams of melted steel are their familiar playthings; yet they are but men.

The "rolling" of these slabs and ingots into rails is a following operation still. The continuous rail is often more than a hundred feet in length, which is cut into three or four rails of thirty feet each, and it goes through every operation that makes it a "T" rail weighing ninety pounds to the yard with the single first heat. There are trains of rolls that will take in a piece of white-hot metal weighing six tons, and send it out in a long sheet three thirty-seconds of an inch thick and nearly ten feet wide. The first steel rails made in this country were made by the Chicago Rolling Mill Company, in May, 1865. Only six rails were then made, and these were laid in the tracks of the Chicago and North Western Railroad. It is said they lasted over ten years. The first nails, or tacks, were made of steel at Bridgewater, Mass., at about the same date.

[Illustration: ROLLING INGOTS.]

Some thirty years ago there were but two Bessemer converters in the United States, and the manufacture of steel did not reach then five hundred tons per annum. In 1890 the product was more than five million tons.

In 1872 the price of steel was one hundred and eighty-six dollars per gross ton. It can be purchased now at varying prices less than thirty dollars per ton. The consumption of seventy millions of people is so great that it is difficult to imagine how so enormous a mass of almost imperishable material can be absorbed, and the latest figures show a consumption greatly in excess of those mentioned as the sum of manufactures.

We turn again for the comparison without which all figures are valueless to the good year 1643, when the "General court" passed a resolve commending the great progress made in the manufacture of iron which they had licensed two years before, and granted the company still further privileges and immunities upon condition that it should furnish the people "with barre iron of all sorts for their use at not exceedynge twenty pounds per ton." We recall the first little piece of hollow ware made in America. We remember how old the old world is said to be and how long the tribes of men have plodded upon it, and then the picture appears of the progress that has grown almost under our eyes. The real Age of Steel began in 1865. It is not yet thirty years old. By comparison we are impressed with the fact that the real history of the metal is compressed into less than half an ordinary lifetime.

THE STORY OF ELECTRICITY

[Illustration: ERIPUIT CAELO FULMEN, SCEPTRUMQUE TYRANNIS.]

There is a sense in which electricity may be said to be the youngest of the sciences. Its modern development has been startling. Its phenomena appear on every hand. It is almost literally true that the lighting has become the servant of man.

But it is also the oldest among modern sciences. Its manifestations have been studied for centuries. So old is its story that it has some of the interest of a mediaeval romance; a romance that is true. Steam is gross, material, understandable, noisy. Its action is entirely comprehensible. The explosives, gunpowder, begriming the nations in all the wars since 1350, nitroglycerine, oxygen and hydrogen in all the forms of their combination, seem to be gross and material, the natural, though ferocious, servants of mankind. But electricity floats ethereal, apart, a subtle essence, shining in the changing splendors of the aurora yet existent in the very paper upon which one writes; mysteriously everywhere; silent, unseen, odorless, untouchable, a power capable of exemplifying the highest majesty of universal nature, or of lighting the faint glow of the fragile insect that flies in the twilight of a summer night. Obedient as it has now been made by the ingenuity of modern man, docile as it may seem, obeying known laws that were discovered, not made, it yet remains shadowy, mysterious, impalpable, intangible, dangerous. It is its own avenger of the daring ingenuity that has controlled it. Touch it, and you die.

Electricity was as existent when the splendid scenes described in Genesis were enacted before the poet's eye as it is now, and was entirely the same. Its very name is old. Before there were men there were trees. Some of these exuded gum, as trees do now, and this gum found a final resting place in the sea, either by being carried thither by the currents of the streams beside which those trees grew, or by the land on which they stood being submerged in some of the ancient changes and convulsions to which the world has been frequently subject. In the lapse of ages this gum, being indestructible in water, became a fossil beneath the waves, and being in later times cast up by storms on the shores of the Baltic and other seas, was found and gathered by men, and being beautiful, finally came to be cut into various forms and used as jewelry. One has but to examine his pipe-stem, or a string of yellow beads, to know it even now. It is amber. The ancient Greeks knew and used it as we do, and without any reference to what we now call "electricity" their name for it was ELEKTRON. The earliest mention of it is by Homer, a poet whose personality is so hidden in the mists of far antiquity that his actual existence as a single person has been doubted, and he mentions it in connection with a necklace made of it.

But very early in human history, at least six hundred years before Christ, this elektron had been found to possess a peculiar property that was imagined to belong to it alone. It mysteriously attracted light bodies to it after it had been rubbed. Thales, the Franklin of his remote time, was the man who is said to have discovered this peculiar and mysterious quality of the yellow gum, and if it be true, to him must be conceded the unwitting discovery of electricity. It was the first step in a science that usurps all the prerogatives of the ancient gods. He recorded his discovery, and was impressed with awe by it, and accounted for the phenomenon he had observed by ascribing to the dull fossil a living soul. That is the unconscious impression still, after twenty-five hundred years have passed since Thales died; that hidden in the heart of electrical phenomena there is a weird sentience; what a Greek would consider something divine and immortal apart from matter. But neither Thales, nor Theophrastus, nor Pliny the elder, nor any ancient, could conceive of a fact but dimly guessed until the day of Franklin; that this secret of the silent amber was also that of the thunder-cloud, that the essence that drew to it a floating filament is also that which rends an oak, that had splintered their temples and statues, and had not spared even the image of Jupiter Tonans himself. The spectral lights which hung upon the masts of the ancient galleys of the Mediterranean were named Castor and Pollux, not electricity. Absolutely no discovery was made, though the religion of ancient Etruria was chiefly the worship of a spirit by them seen, but unknown; to us electrical science; a science chained, yet really unknown and still feared though chained. It is the story of this servitude only that is capable of being told, and the first weak bands were a hundred and forty-six years in forging; from the Englishman Gilbert's "*De Magnete*," to Franklin's Kite.

During all this time, and to a great degree long after, electricity was a scientific toy. Experiences in the sparkling of the fur of cats, the knowledge that there were fishes that possessed a mysterious paralyzing power, and various common phenomena all attributable to some unknown common cause, did not greatly increase the sum of actual knowledge of the subject. There was no divination of what the future would bring, and not the least conception of actual and impending possibilities. When, finally, the greatest thinkers of their times began to investigate; when Boyle began to experiment, and even the transcendent genius of Newton stooped to enquiry; from the days of those giants down to those of the American provincial postmaster, Benjamin Franklin, a period of some seventy years, almost all the knowledge obtained was only useful in indicating how to experiment still further. So small was the knowledge, so aimless the long experimenting, that the discovery that not amber only, but other substances as well, possessed the electric quality when rubbed, was a notable advance in knowledge. Later, in 1792, it was found by Gray that certain substances possessed the power of carrying; "conducting" as we now term it; the mysterious fluid from one substance to another; from place to place. This discovery constituted an actual epoch in the history of the science, and justly, since this small beginning with a wet string and a cylinder of glass or a globe of sulphur was the first unwitting illustration of the net-work of wires now hanging all over the world. The next step was to find that all substances were not alike in a power to conduct a current; *i.e.*, that there were "conductors" and "non-conductors," and all varying grades and powers between. The next discovery was that there were, as was then imagined, several kinds of electricity. This conclusion was incorrect, and its use was to lead at last to the discovery, by Franklin, that the many kinds were but two, and even these not kinds, but qualities, present always in the unchanging essence that is everywhere, and which are known to us now by the names that Franklin gave them; the *positive* and *negative* currents; one always present with the other, and in every phenomenon known to electrical science.

Probably the first machine ever contrived for producing an electric current was made by a monk, a Scotch Benedictine named Gordon who lived at Erfurt, in Saxony. I shall have occasion, hereafter, to describe other machines for the same purpose, and this first contrivance is of interest by comparison. It was a cylinder of glass about eight inches long, with a wooden shaft in the center, the ends of which were passed through holes in side-pieces, and it is said to have been operated by winding a string around the shaft and drawing the ends of the string back and forth alternately.

[Illustration: THE FIRST ELECTRICAL MACHINE.]

The Franklinic machine, the modern glass disc fitted with combs, rubbers, bands and cranks, is nothing more in principle or manner of action than the first crude arrangement of the monk of Erfurt.

All these experiments, and all that for many years followed, were made in electricity produced by friction; by rubbing some body like glass, sulphur or rosin. Many men took part in producing effects that were almost meaningless to them—the preliminaries to final results for us. Improved electrical machines were made, all seeming childish and inadequate now, and all wonderful in their day. There is a long list of immortal names connected with the slow development of the science, and among their experiments the seventeenth century passed away. Dufaye and the Abbe Nollet worked together about 1730, and mutually surprised each other daily. Guericke, better known as the inventor of the air-pump, made a sulphur-ball machine, often claimed to have been the first. Hawkesbee constructed a glass machine that was an improvement over that of Guericke. Stephen Gray unfolded the leading principles of the science, but without any understanding of their results as we now understand them. The next advance was made in finding a way to hold some of the electricity when gathered, and the toy which we know as the Leyden Jar surprised the scientific world. Its inventor, Professor Muschenbrock, wrote an account of it to Réaumur, and lacks language to express the terror into which his own experiments had thrown him. He had unwittingly accumulated, and had accidentally discharged, and had, for the first time in human experience, felt something of the shock the modern lineman dreads because it means death. He had toiled until he held the baleful genie in a glass vessel partially filled with water, and the sprite could not be seen. Accidentally he made a connection between the two surfaces of the jar, and declared that he did not recover from the experience for two days, and that nothing could induce him to repeat it. He had been touched by the lightning, and had not known it. [Footnote: The Leyden Jar has little place in the usefulness of modern electricity, and has no relationship with the modern so-called "Storage" Battery.]

Then began the fakerism which attached itself to the science of electricity, and that has only measurably abandoned it in very late times. Itinerant electricians began to infest the cities of Europe, claiming medicinal and almost supernatural virtues for the mysterious shock of the Leyden Vial, and showing to gaping multitudes the quick and flashing blue spark which was, though no man knew it then, a miniature imitation of the bolt of heaven. That fact, verging as closely upon the sublimest power of nature as a man may venture to and live, was not even suspected until Franklin had invented a battery of such jars, and had performed hundreds of experiments therewith that finally established in his acute, though prosaic, mind the identity of his puny spark with that terrific flash that, until that time, had been regarded by all mankind as a direct and intentional expression of the power of Almighty God.

Thus Franklin came into the field. He was an investigator who brought to his aid a singular capacity possessed by the very few; the capacity for an unbiased looking for the hidden reasons of things. There was no field too sacred or too old for his prying investigations and his private conclusions. He was, as much as any man ever is, an original thinker. He knew of all the electrical experiments of others, and they produced in his mind conclusions distinctly his own. He was, upon topics pertaining to the field of reason, experience and common sense, the clearest and most vigorous writer of his time save one, and such conclusions as he arrived at he knew how to promulgate and explain. All that Franklin discovered would but add to the tedium of the subject of electricity now, but from his time definitely dates the knowledge that of electricity, in all its developments, there is really but one kind, though for convenience sake we may commonly speak of two, or even more. He first gave the names by which they are still known to the two qualities of one current; a name of convenience only. He knew first a fact that still puzzles inquiry, and is still largely unknown—that electricity is not *created*, produced, manufactured, by any human means, and that all we may do, then or now, is to gather it from its measureless diffusion in the air, the world, or the spaces of the wide creation, and that, like "heat" and "cold," it is a relative term. He demonstrated that any body which has electricity gives it to any other body that has at the moment less. Before he had actually tried that celebrated experiment which is alone sufficient to give him place among the immortals, he had declared the theory upon which he made it to be true, and by reasoning, in an age that but dimly understood the force and conditions of inductive reason, had proved that lightning is but an electric spark. It seems hardly necessary to add that his theories were ridiculed by the most intelligent scientists of his time, and scoffed at even by the countrymen of Newton and Davy, the members of the Royal Society of England. Franklin was a provincial American, and had, in other fields than electricity, troubled the British placidity.

[Illustration: B. FRANKLIN]

Only one of these, a man named Collinson, saw any value in these researches of the provincial in the wilds of America. He published Franklin's letters to him. Buffon read them, and persuaded a friend to translate them into French. They were translated afterwards into many languages, and when in his isolation he did not even know it, the obscure printer, the country postmaster who kept his official accounts with his own hands, was the bearer of a famous name. He was assailed by the Nollet previously mentioned, and by a party of French philosophers, yet there arose, in his absence and without his knowledge, a party who called themselves distinctively "Franklinists."

Then came the personal test of the truth of these theories that had been promulgated over Europe in the name of the unknown American. He was then forty-five years old, successful in his walk and well-known in his immediate locality, but by no means as prominent or famous among his neighbors as he was in Europe. He was not so fertile in resources as to be in any sense inspired, and had privately waited for the finishing of a certain spire in the little town of Philadelphia so that he might use it to get nearer to the clouds to demonstrate his theory of lightning. It was in June, 1752, that this great exemplar of the genius of common-sense descended to the trial of the experiment that was the simplest and the most ordinary and the most sublime; the commonest in conception and means yet the most famous in results; ever tried by man. He had grown impatient of delay in the matter of the spire, and hastily, as by a sudden thought, made a kite. It was merely a silk handkerchief whose four corners were attached to the points of two crossed sticks. It was only the idea that was great; the means were infantile. A thunder shower came over, and in an interval between sprinklings he took with him his son, and went by back ways and alleys to a shed in an open field. The two raised the kite as boys did then and do now, and stood within the shelter. There was a hempen string, and on this, next his hand, he had tied a bit of ribbon and an ordinary iron key. A cloud passed over without any indications of anything whatever. But it began to rain, and as the string became wet he noticed that the loose filaments were standing out from it, as he had often seen them do in his experiments with the electrical machine. He drew a spark from the key with his finger, and finally charged a Leyden jar from this key, and performed all the then known proof-experiments with the lightning drawn from heaven.

It is manifest that the slightest indication of the presence of the current in the string was sufficient to have demonstrated the fact which Franklin sought to fix. But it would have been insufficient to the general mind. The demonstration required was absolute. Even among scientists of the first class less was then known about electricity and its phenomena, and the causes of them, than now is known by every child who has gone to school. No estimate of the boldness and value of Franklin's renowned experiment can be made without a full appreciation of his times and surroundings. He demonstrated that which was undreamed before, and is undoubted now. The wonders of one age have been the toys and tools of the next through the entire history of mankind. The meaning of the demonstration was deep; its results were lasting The experimenters thereafter worked with a knowledge that their investigations must, in a sense, include the universe. Perhaps the obscure man who had toyed with the lightnings himself but vaguely understood the real meaning of his temerity. For he had, as usual, an intensely practical purpose in view. He wished to find a way of "drawing from the heavens their lightnings, and conducting them harmless to the earth." He was the first inventor of a practical machine, for a useful purpose, with which electricity had to do. That machine was the lightning-rod. Whatever its purpose, mankind will not forget the simple greatness of the act. At this writing the statue of Franklin stands looking upward at the sky, a key in his extended hand, in the portico of a palace which contains the completest and most beautiful display of electrical appliances that was ever brought together, at the dawn of that Age of Electricity which will be noon with us within one decade. The science and art of the civilized world are gathered about him, and on the frieze above his head shines, in gold letters, that sentence which is a poem in a single line. "ERIPUIT CAELO FULMEN, SCEPTRUMQUE TYRANNIS." [Footnote: "He snatched the lightning from heaven, and the sceptre from tyrants."]

* * * * *

THE MAN FRANKLIN.—Benjamin Franklin was born at Boston, Mass., Jan. 17th, 1706. His father was a chandler, a trade not now known by that term, meaning a maker of soaps and candles. Benjamin was the fifteenth of a family of seventeen children. He was so much of the same material with other boys that it was his notion to go to sea, and to keep him from doing so he was apprenticed to his brother, who was a printer. To be apprenticed then was to be absolutely indentured; to belong to the master for a term of years. Strangely enough, the boy who wanted to be a sailor was a reader and student, captivated by the style of the *Spectator*, a model he assiduously cultivated in his own extensive writings afterwards. He was not assisted in his studies, and all he ever knew of mathematics he taught himself. Being addicted to literature by natural proclivity he inserted his own articles in his brother's newspaper, and these being very favorably commented upon by the local public, or at least noticed and talked about, his authorship of them was discovered, and this led to a quarrel between the two brothers. Nevertheless, when James, the elder brother, was imprisoned for alleged seditious articles printed by him, the paper was for a time issued in young Benjamin's name. But the quarrel continued, the boy was imposed upon by his master, and brother, as naturally as might have been expected under the circumstances of the younger having the monopoly of all the intellectual ability that existed between the two, and in 1723, being then only seventeen, he broke his indentures, a heinous offense in those times, and ran away, first to New York and then to Philadelphia, where he found employment as a journeyman printer. He had attained a skill in the business not usual at the time.

The boy had, up to this time, read everything that came into his hands. A book of any kind had a charm for him. His father observing this had intended him for the ministry, that being the natural drift of a pious father's mind in the time of Franklin's youth, when he discovered any inclination to books on the part of a son. But, later, he would neglect the devotions of the Sabbath if he had found a book, notwithstanding the piety of his family. Sometimes he distressed them further by neglecting his meals, or sitting up at night, for the same reason. There is no question that young Franklin was a member of that extensive fraternity now known as "cranks." [Footnote: Most people, then and now, can point to people of their acquaintance whom they hold in regard as originals or eccentrics. It is a somewhat dubious title for respect, even with us who are reckoned so eccentric a nation. And yet all the great inventions which have done so much for civilization have been discovered by eccentrics—that is, by men who stepped out of the common groove; who differed more or less from other men in their habits and ideals.] He read a book advocating exclusive subsistence upon a vegetable diet and immediately adopted the idea, remaining a disciple of vegetarianism for several years. But there is another reason hinted. He saved money by the vegetable scheme, and when his printer's lunch had consisted of "biscuits (crackers) and water" for some days, he had saved money enough to buy a new book.

This young printer, who, at school, in the little time he attended one, had "failed entirely in mathematics," could assimilate "Locke on the Understanding," and appreciate a translation of the Memorabilia of Xenophon. Even after his study of this latter book he had a fondness for the calm reasoning of Socrates, and wished to imitate him in his manner of reasoning and moralizing. There is no question but that the great heathen had his influence across the abyss of time upon the mind of a young American destined also to fill, in many respects, the foremost place in his country's history. There was one, at least, who had no premonition of this. His brother chastised him before he had been imprisoned, and after he had begun to attract attention as a writer in one of the only two newspapers then printed in America, and beat him again after he was released, having meantime been vigorously defended by his apprentice editorially while he languished. To have beaten Benjamin Franklin with a stick, when he was seventeen years old, seems an absurd anticlimax in American history. But it is true, and when the young man ran away there was still another odd episode in a great career.

Upon his first arrival in Philadelphia as a runaway apprentice, with one piece of money in his pocket, occurs the one gleam of romance in Franklin's seemingly Socratic life. He says he walked in Market Street with a baker's loaf under each arm, with all his shirts and stockings bulging in his pockets, and eating a third piece of bread as he walked, and this on a Sunday morning. Under these circumstances he met his future wife, and he seems to have remembered her when next he met her, and to have been unusually prepossessed with her, because on the first occasion she had laughed at him going by. He was one of those whose sense of humor bears them through many difficulties, and who are even attracted by that sense in others. He was, at this period, absurd without question. Having eaten all the bread he could, and bestowed the remainder upon another voyager, he drank out of the Delaware and went to church; that is, he sat down upon a bench in a Quaker meetinghouse and went to sleep, and was admonished thence by one of the brethren at the end of the service.

Franklin had, in the time of his youth, the usual experiences in business. He made a journey to London upon promises of great advancement in business, and was entirely disappointed, and worked at his trade in London. Afterwards, during the return voyage to America, he kept a journal, and wrote those celebrated maxims for his own guidance that are so often quoted. The first of these is the gem of the collection: "I resolve to be extremely frugal for some time, until I pay what I owe." A second resolve is scarcely less deserving of imitation, for it declares it to be his intention "to speak all the good I know of everybody." It must be observed that Franklin was afterwards the great maximist of his age, and that his life was devoted to the acquisition of worldly wisdom. In his body of philosophy there is included no word of confidence in the condemnation of offenses by the act or virtue of another, no promise of, or reference to, the rewards of futurity.

When about twenty-one years of age, we find this old young man tired of a drifting life and many projects, and desiring to adopt some occupation permanently. He had courted the girl who had laughed at him, and then gone to England and forgotten her. She had meantime married another man, and was now a widow. In 1730 he married her. Meantime, entering into the printing business on his own account, he often trundled his paper along the streets in a wheelbarrow, and was intensely occupied with his affairs. His acquisitive mind was never idle, and in 1732 he began the publication of the celebrated "Poor Richard's Almanac." This was among the most successful of all American publications, was continued for twenty-five years, and in the last issue, in 1757, he collected the principal matter of all preceding numbers, and the issue was extensively republished in Great Britain, was translated into several foreign languages, and had a world-wide circulation. He was also the publisher of a newspaper, *The Pennsylvania Gazette*, which was successful and brought him into high consideration as a leader of public opinion in times which were beginning to be troubled by the questions that finally brought about a separation from the mother country.

Time and space would fail in anything like a detailed account of the life of this remarkable man. His only son, the boy who was with him at the flying of the kite, was an illegitimate child, and it is a remarkable instance of unlikeness that this only son became a royalist governor of New Jersey, was never an American in feeling, and removed to England and died there. The sum of Franklin's life is that he was a statesman, a financier of remarkable ability, a skillful diplomat, a law-maker, a powerful and felicitous writer though without imagination or the literary instinct, and a controversialist who seldom, if ever, met his equal. He was always a printer, and at no period of his great career did he lose his affection for the useful arts and common interests of mankind. He is the founder of the

American Philosophical Society, and of a college which grew into the present University of Pennsylvania. To him is due the origin of a great hospital which is still doing beneficent work. He raised, and caused to be disciplined, ten thousand men for the defense of the country. He was a successful publisher of the literature of the common people, yet a literature that was renowned. He could turn his attention to the improvement of chimneys, and invented a stove still in use, and still bearing his name as the author of its principle. [Footnote: The stove was not used in Franklin's time to any extent. The "Franklin Stove" was a fireplace so far as the advantages were concerned, such as ventilation and the pleasure of an open fire. But it also radiated heat from the back and sides as well as the front, and was intended to sit further out into a room; to be both fireplace and stove.] He organized the postal system of the United States before the Union existed. He was a signer of the Declaration of Independence. He sailed as commissioner to France at the age of seventy-one, and gave all his money to his country on the eve of his departure, yet died wealthy for his time. Serene, even-tempered, philosophical, he was yet far-seeing, care-taking, sagacious, and intensely industrious. He acquired a knowledge of the Italian and Spanish languages, and was a proficient French speaker and writer. He possessed, in an extraordinary degree, the power of gaining the regard, even the affection, of his fellow-men. He was even a competent musician, mastering every subject to which his attention was turned; and province-born and reared in the business of melting tallow and setting types, without collegiate education, he shone in association with the men and women who had place in the most brilliant epoch of French intellectual history. At fourscore years he performed the work that would have exhausted a man of forty, and at the same time wrote, for mere amusement, sketches such as the "Dialogue between Franklin and the Gout," and added, with the cool philosophy of all his life still lingering about his closing hours: "When I consider how many terrible diseases the human body is liable to, I think myself well off that I have only three incurable ones, the gout, the stone, and old age."

[Illustration: THE FRANKLIN STOVE.]

* * * * *

After Franklin, electrical experiments went on with varying results, confined within what now seems to have been a very narrow field, until 1790. The great facts outside of the startling disclosure made by Franklin's experiments remained unknown. It was another forty years of amused and interested playing with a scientific toy. But in that year the key to the *utility* of electricity was found by one Galvani. He was not an electrician at all, but a professor of anatomy in the university of Bologna. It may be mentioned in passing that he never knew the weight or purport of his own discovery, and died supposing and insisting that the electric fluid he fancied he had discovered had its origin in the animal tissues. Misapprehending all, he was yet unconsciously the first experimenter in what we, for convenience, designate *dynamic* electricity. He knew only of *animal* electricity, and called it by that name; a misnomer and a mistake of fact, and the cause of an early scientific quarrel the promoting of which was the actual reason of the advance that was made in the science following his accidental and enormously important discovery.

There are many stories of the details of the ordinarily entirely unimportant circumstances that led to *Galvanism* and the *Galvanic Battery*. Volta actually made this battery, then known as the Voltaic Pile, but he made it because of Galvani's discovery. The reader is requested to bear these names in mind; Galvani and Volta. They have a unique claim upon us. With others that will follow, they have descended to all posterity in the immortal nomenclature of the science of electricity. It is through the accidental discovery of the plodding demonstrator of anatomy in a medical college, a man who died at last in poverty and in ignorance of the meaning of his own work, that we have now the vast web of telegraph and telephone wires that hangs above the paths of men in every civilized country, and the cables that lie in the ooze of the oceans from continent to continent. His discovery was the result of one of the commonest incidents of domestic life. Variously described by various writers, the actual circumstance seems reducible to this.

In Galvani's kitchen there was an iron railing, and immediately above the railing some copper hooks, used for the purpose of hanging thereon uncooked meats. His wife was an invalid, and wishing to tempt her appetite he had prepared a frog by skinning it, and had hung it upon one of the copper hooks. The only use intended to be asked of this renowned batrachian was the making of a little broth. Another part of the skinned anatomy touched the iron rail below, and the anatomist observed that this casual contact produced a convulsive twitching of the dead reptile's legs. He groped about this fact for many years. He fancied he had discovered the principle of life. He made the phenomenon to hang upon the facts clustering about his own profession, familiar to him, and about which it was natural for him to think. He promulgated theories about it that are all now absurd, however tenable then. His was an instance of how the fatuities of men in all the fields of science, faith or morals, have often led to results as extraordinary as they have been unexpected. That he died in poverty in 1798 is a mere human fact. That in this life he never knew is merely another. It is but a part of that sadness that, through life, and, indeed, through all history, hangs over the earthly limitations of the immortal mind.

Volta, his contemporary and countryman, finally solved the problem as to the reason why. and made that "Voltaic Pile" which came to be our modern "battery." Acting upon the hint given by Galvani's accident, this pile was made of thin sheets of metal, say of copper and zinc, laid in series one above the other, with a piece of cloth wet with dilute acid interposed between each sheet and the next. The sheets were connected at the edges in pairs, a sheet of zinc to a sheet of copper, and the pile began with a sheet of one metal and ended with one of the other. It is to be noted that a single pair would have produced the same result as a hundred pairs, only more feebly. A single large pair is, indeed, the modern electric battery of one cell. The beginning and the ending sheets of the Voltaic pile were connected by a wire, through which the current passed. We, in our commonest industrial battery, use the two pieces of metal with the fluid between. The metals are usually copper and zinc, and the fluid is water in which is dissolved sulphate of copper. The wire connection we make hundreds of miles long, and over this wire passes the current. If we part this wire the current ceases. If we join it again we instantly renew it. There are many forms of this battery. The two metals, the *electrodes*, are not necessarily zinc and copper and no others. The acidulated fluid is not invariably water with sulphate of copper dissolved in it. Yet in all modifications the same thing is done in essentially the same way, and the Voltaic pile, and a little back of that Galvani's frog, is the secret of the telegraph, the telephone, the telautograph, the cable message. In the case of Galvani's frog, the fluids of the recently killed body furnished the liquid containing the acid, the copper hook and the iron railing furnished the dissimilar metals, and the nerves and muscles of the frog's body, connecting the two metals, furnished the wire. They were as good as Franklin's wet string was. The effect of the passage of a current of electricity through a muscle is to cause it to spasmodically contract, as everyone knows who has held the metallic handles of an ordinary small battery. Many years passed before the mystery that has long been plain was solved by acute minds. Galvani thought he saw the electric quality *in the tissues of the* frog. Volta came to see them as produced *by chemical action upon two dissimilar metals*. The first could not maintain his theories against facts that became apparent in the course of the investigations of several years, yet he asserted them with all the pertinacious conservatism of his profession, which it has required ages to wear away, and died poor and unhonored. The other became a nobleman and a senator, and wore medals and honors. It is a world in which success alone is seen, and in which it may be truthfully said that the contortions of an eviscerated and unconscious frog upon a casual hook were the not very remote cause of the greatest advancements and discoveries of modern civilization.

Yet the mystery is not yet entirely explained. In the study of electricity we are accustomed to accept demonstrated facts as we find them. When it is asked *how* a battery acts, what produces the mysterious current, the only answer that can now be given is that it is *by the conversion of the energy of chemical affinity into the energy of electrical vibrations*. Many mixtures produce heat. The explanation can be no clearer than that for electricity. Electricity and heat are both *forms of energy*, and, indeed, are so similar that one is almost synonymous with the other. The enquiry into the original sources of energy, latent but present always, will, when finally answered, give us an insight into mysteries that we can only now infer are reserved for that hereafter, here or elsewhere, which it is part of our nature to believe in and hope for. The theory of electrical vibrations is explained elsewhere as the only tenable one by which to account for electrical action. One may also ask how fire burns, or, rather, why a burning produces what we call "heat," and the actual question cannot be answered. The action of fire in consuming fuel, and the action of chemicals in consuming metals, are similar actions. They each result in the production of a new form of energy, and of energy in the form of vibrations. In the action of fire the vibrations are irregular and spasmodic; in electricity they are controlled by a certain rhythm or regularity. Between heat and electricity there is apparently only this difference, and they are so similar, and one is so readily converted into the other, that it is a current scientific theory that one is only a modified form of the other. Many acute minds have reflected upon the problem of how to convert the latent energy of coal into the energy of electricity without the interposition of the steam engine and machinery. There apparently exist reasons why the problem will never be solved. There is no intelligence equal to answering the question as to precisely where the heat came from, or how it came, that instantly results upon the striking of a common match. It was *evolved* through friction. The means were necessary. Friction, or its precise equivalent in energy, must occur. The result is as strange, and in the same manner strange, as any of the phenomena of electricity. Precisely here, in the beginning of the study of these phenomena, the student should be warned that an attitude of wonder or of awe is not one of enquiry. The demonstrations of electricity are startling chiefly for three reasons: newness, silence, and inconceivable rapidity of action. Let one hold a wire in one's hand six or eight inches from the end, and then insert that end into the flame of a gas-jet. It is as old as human experience that that part of the wire which is not in the flame finally grows hot, and burns one's fingers. A change has taken place in the molecules of the wire that is not visible, is noiseless, and that has *traveled along the wire*. It excites neither wonder nor remark. No one asks the reason why. Yet it cannot be explained except by some theory more or less tenable, and the phenomenon, in kind though not in degree, is as unaccountable as anything in the magic of electricity. In a true sense there is, nothing supernatural, or even wonderful, in all the vast universe of law. If we would learn the facts in regard to anything, it must be after we have passed the stage of wonder or of reverence in respect to it. That which was the "Voice of God"—as truly, in a sense, it was and is—until Franklin's day, has since been a concussion of the air, an echo among the clouds, the passage of an electric discharge. It is the first lesson for all those who would understand.

The time had now come when that which had seemed a lawless wonder should have its laws investigated, formulated and explained. A man named Coulomb, a Frenchman, is the author of a system of measurements of the electric current, and he it was who discovered that the action of electricity varies, not with the distance, but, like gravity, *in the inverse ratio of the square of the distance*. Coulomb was the maker of the first instrument for measuring a current, which was known as the *torsion balance*. The results of his practical investigations made easier the practical application of electrical power as we now use it, though he foresaw nothing of that application; and the engineer of to-day applies his laws, and those of his fellow scientists, as those which do not fail. Volta was one of these, and he also furnished, as will hereafter be seen, a name for one of the units of electrical measurement.

Both Galvani and Volta passed into shadow, when, in 1820, Professor H. C. Oersted, of Copenhagen, discovered the law upon which were afterwards slowly built the electrical appliances of modern life. It was the great principle of INDUCTION. The student of electricity may begin here if he desires to study only results, and is not interested in effects, causes, and the pains and toils which led to those results. The term may seem obscure, and is, doubtless, as a name, the result of a sudden idea; but upon induction and its laws the simplest as well as the most complicated of our modern electrical appliances depend for a reason for action. Its discovery set Ampère to work. They had all imagined previously that there was some connection between electricity and magnetism, and it was this idea that instigated the investigations of Ampere. It was imagined that the phenomena of electricity were to be explained by magnetism. This was not untrue, but it was only a part of the truth. Ampere proved that *magnetism could also readily be produced by a current of electricity*. From this idea, practically carried out, grew the ELECTRO MAGNET, and to Ampère we are indebted for the actual discovery of the elementary principles of what we now call electrodynamics, or dynamic electricity, [Footnote: In all science there is a continual going back to the past for a means of expression for things whose application is most modern. *Dynamic*; DYNAMO, is the

Greek word for power; to be able. Once established, these names are seldom abandoned. There is no more reason for calling our electrical power-producing machine a "Dynamo" than there would be in so designating a steam engine or a water-wheel. But, a term of general significance if used at all, it has come to be the special designation of that one machine. It is brief, easily said, and to the point, but is in no way necessarily connected with *electrical* power distinctively.] in which are included the Dynamo, and its twin and indispensable, the Motor. Ampère is also the author of the *molecular theory*, by which alone, with our present knowledge, can the action of electricity be explained in connection with the iron core which is made a magnet by the current, and left again a mere piece of iron when the current is interrupted. Ten years later Faraday explained and applied the laws of Induction, basing them upon the demonstrations of Ampère. The use of a core of soft iron, magnetized by the passage of a current through a helix of wire wrapping it as the thread does a spool, is the indispensable feature, in some form meaning the same thing, with the same results, in all machines that are given movement to by an electric current. This is the electro-magnet. It is made a magnet not by actual contact, or by being made the conductor of a current, but by being placed in the "electrical field" and temporarily magnetized by induction.

Faraday began his brilliant series of experiments in 1831. To express briefly the laws of action under which he worked, he wrote the celebrated statement of the Law of Magnetic Force. He proved that the current developed by induction is the same in all its qualities with other currents, and, indeed, demonstrated Franklin's theory that all electricity is the same; that, as to *kind*, there is but one. All electrical action is now viewed from the Faradic position.

The story of electricity, as men studied it in the primary school of the science, ends where Faraday began. Under the immutable laws he discovered and formulated we now enter the field of result, of action, of commercial interest and value. We might better say the field of usefulness, since commercial value is but another expression for usefulness. A revolution has been wrought in all the ways and thoughts of men since a date which a man less than sixty years old can recall. The laws under which the miracle has been wrought existed from all eternity. They were discovered but yesterday. Progress, the destiny of man, has kept pace in other fields. We live our time in our predestined day, learning and knowing, like grown-up children, what we may. In a future whose distance we may not even guess, the children of men shall reap the full fruition of the prophesy that has grown old in waiting, and "shall be as gods, knowing good from evil."

MODERN ELECTRICITY

CHAPTER I.

Electricity, in all its visible exhibitions, has certain unvarying qualities. Some of these have been mentioned in the preceding chapter. Others will appear in what is now to follow. These qualities or habits, invariable and unchangeable, are, briefly:

(1) It has the unique power of drawing, "attracting" other objects at a distance.

(2) For all human uses it is instantaneous in action, through a conductor, at any distance. A current might be sent around the world while the clock ticked twice.

(3) It has the power of decomposing chemicals (Electrolysis), and it should be remembered that even water is a chemical, and that substances composed of one pure organic material are very rare.

(4) It is readily convertible into heat in a wire or other conductor.

These four qualities render its modern uses possible, and should be remembered in connection with what is presently to be explained.

These uses are, in application, the most startling in the entire history of civilization. They have come about, and their applications have been made effective, within twenty years, and largely within ten. This subtlest and most elusive essence in nature, not even now entirely understood, is a part of common life. Some years ago we began to spell our thoughts to our fellow-men across land and sea with dots and dashes. Within the memory of the present high school boy we began to talk with each other across the miles. Now there is no reason why we shall not begin to write to each other letters of which the originals shall never leave our hands, yet which shall stand written in a distant place in our own characters, indisputably signed by us with our own names. We apparently produce out of nothing but the whirling of a huge bobbin of wire any power we may wish, and send it over a thin wire to where we wish to use it, though every adult can remember when the difficulty of distance, in the propelling of machinery, was thought to have been solved to the satisfaction of every reasonable man by the making of wire cables that would transmit power between grooved wheels a distance of some hundreds of feet. We turn night into day with the glow of lamps that burn without flame, and almost without heat, whose mysterious glow is fed from some distant place, that hang in clusters, banners, letters, in city streets, and that glow like new stars along the treeless prairie horizon where thirty years ago even the beginnings of civilization were unknown. Yet the mysterious agent has not changed. It is as it was when creation began to shape itself out of chaos and the abyss. Men have changed in their ability to reason, to deduce, to discover, and to construct. To know has become a part of the sum of life; to understand or to abandon is the rule. When the ages of tradition, of assertion without the necessity for proof, of content with all that was and was right or true because it was a standard fixed, went by, the age not necessarily of steam, or of steel, or of electricity, but the age of thought, came in. Some of the results of this thought, in one of the most prominent of its departments, I shall attempt to describe.

A wire is the usual concomitant in all electrical phenomena. It is almost the universally used conductor of the current. In most cases it is of copper, as pure as it can be made in the ordinary course of manufacture. There are other metals that conduct an electrical current even better than copper does, but they happen to be expensive ones, such as silver. The usual telegraph-line is efficient with only iron wire.

We habitually use the words "conductor" and "conduct" in reference to the electric current. A definition of that common term may be useful. It is a relative one. *A conductor is any substance whose atoms, or molecules, have the power of conveying to each other quickly their electricities.* Before the common use of electricity we were accustomed to commonly speak of conductors of heat; good, or poor. The same meaning is intended in speaking of conductors of electricity. *Non-conductors are those whose molecules only acquire this power under great pressure.* Electricity always takes the *easiest* road, not necessarily the shortest. This is the path that electricians call that of "least resistance." There are no absolutely perfect conductors, and there are no substances that may be called absolutely non-conductors. A non-conductor is simply a reluctant, an excessively slow, conductor. In all electrical operations we look first for these two essentials: a good conductor and a good non-conductor. We want the latter as supports and attachments for the first. If we undertake to convey water in a pipe we do not wish the pipe to leak. In conveying electricity upon a wire we have a little leak wherever we allow any other conductor to come too near, or to touch, the wire carrying the current. These little electrical leaks constantly exist. All nature is in a conspiracy to take it wherever it can find it, and from everything which at the moment has more than some other has, or more than its share with reference to the air and the world, of the mysterious essence that is in varying quantities everywhere. Glass is the usual non-conductor in daily use. A glance at the telegraph poles will explain all that has just been said. Water in large quantity or widely diffused is a fair conductor. Therefore, the glass insulators on the telegraph-poles are cup-shaped usually on the under side where the pin that holds them is inserted, so that the rain may not actually wet this pin, and thus make a water-connection between the wire, glass, pin, pole and ground.

We are accustomed to things that are subject to the law of gravity. Water will run through a pipe that slants downward. It will pass through a pipe that slants upward only by being pushed. But electricity, in its far journeys over wires, is not subject to gravity. It goes indifferently in any direction, asking only a conductor to carry it. There is also a trait called *inertia*; that property of all matter by which it tends when at rest to remain so, and when in motion to continue in motion, which we meet at every step we take in the material world. Electricity is again an exception. It knows neither gravity, nor inertia, nor material volume, nor space. It cannot be contained or weighed. Nothing holds it in any ordinary sense. It is difficult to express in words the peculiar qualities that caused the early experimenters to believe it had a soul. It is never idle, and in its ceaseless journeyings it makes choice of its path by a conclusion that is unerring and instantaneous.

We find that it is the constant endeavor of electricity to *equalize its quantities and its two qualities, in all substances that are near it that are capable of containing it*. To this end, seemingly by definite intention, it is found on the outsides of things containing it. It gathers on the surfaces of all conductors. If there are knobs or points it will be found in them, ready to leap off. When any electrified body is approached by a conductor, the fluid will gather on the side where the approach is made. If in any conductor the current is weak, very little of it, if any, will go off into the conductor before actual contact is made. If it is strong, it will often leap across the space with a spark. One body may be charged with positive, and another with negative, electricity. There is then a disposition to equalize that cannot be easily repressed. The positive and the negative will assume their dual functions, their existence together, in spite of obstacles. So as to quantity. That which has most cannot be restrained from imparting to that which has less. The demonstration of these facts belongs to the field of experimental, or laboratory, electricity. The most common of the visible experiments is on a vast scale. It is the thunderstorm. Mother Earth is the great depository of the fluid. The heavy clouds, as they gather, are likewise full. Across the space that lies between the exchange takes place — the lightning-flash.

In the preceding chapter I have hastily alluded to the phenomenon known as the key to electricity as a utilitarian science; a means of material usefulness. These uses are all made possible under the laws of what we term INDUCTION. To comprehend this remarkable feature of electric action, it must first be understood that all electrical phenomena occur in what has been termed an "*Electrical Field*" This field may be illustrated simply. A wire through which a current is passing *is always surrounded by a region of attractive force*. It is scientifically imagined to exist in the form of rings around the wire. In this field lie what are termed "lines of force." The law as stated is that the lines in which the magnetism produced by electricity acts *are always at right angles with the direction in which the current is passing*. Let us put this in ordinary phrase, and say that in a wire through which a current is passing there is a magnetic attraction, and that the "pull" is always *straight toward the wire*. This magnetism in a wire, when it is doubled up and multiplied sufficiently, has strong powers of attraction. This multiplying is accomplished by winding the wire into a compact coil and passing a current through it. If one should wind insulated wire around a core, or cylinder, and should then pull out the cylinder and attach the two ends of the wire to the opposite poles of a battery, when the current passed through the coil the hollow interior of it would be a strong magnetic field. The air inside might be said to be a magnet, though if there were no air there, and the coil were under the exhausted receiver of an air-pump, the effect would be the same, and the *vacuum* would be magnetized. A piece of iron inserted where the core was, would instantly become a magnet, and when the insulated wire is wound around a soft iron core, and the core is left in place, we have at once what is known as an *Electro-Magnet*.

The wire windings of an electro-magnet are always insulated; wound with a non-conductor, like silk or cotton; so that the coils may not touch each other in the winding and thus permit the current to run off through contact by the easiest way, and cut across and leave most of the coil without a current. For it may as well be stated now that no matter how good a conductor a wire may be, two qualities of it cause what is called *"resistance"* — the current does not pass so easily. These two qualities are *thinness* and *length*. The current will not traverse all the length of a long coil if it can pass straight through the same mass, and it is made to go the long way *by keeping the wires from touching each other* — preventing "contact," and lessening the opportunity to jump off which electricity is always looking for.

When this coil is wound in layers, like the thread upon a spool, it increases the intensity of the magnetism in the core by as many times as there are coils, up to a certain point. If the core is merely soft iron, and not steel, it becomes magnetized instantly, as stated, and will draw another piece of iron to it with a snap, and hold it there as long as there is a current passing through the coil. But as instantly, when the current is stopped, this soft iron core ceases to be a magnet, and becomes as it was before — an inert and ordinary piece of iron. What has just been described is always, in some form, one of the indispensable parts of the electromagnetic machines used in industrial electricity, and in all of them except the appliances of electric lighting, and even in that case it is indispensable in producing the current which consumes the points of the carbon, or heats the filament to a white glow. The current may traverse the wire for a hundred miles to reach this little coil. But, instantly, at a touch a hundred miles away that forms a contact, there is a continuous "circuit;" the core becomes a magnet, and the piece of iron near it is drawn suddenly to it. Remove the distant finger from the button, the contact is broken, and the piece of iron immediately falls away again. It is the wonder of *the production of instant movement at any distance, without any movement of any connecting part*. It is a mysterious and incredible transmission of force not included among human possibilities forty years ago. It is now common, old, familiar. Conceive of its possibilities, of its annihilation of time and space, of its distant control, and of that which it is made to mean and represent in the spelled-out words of language, and it still remains one of the wonders of the world: the Electric Telegraph.

* * * * *

MAGNETS AND MAGNETISM.—Having described a magnet that is made and unmade at will, it may be appropriate to describe magnets generally. The ordinary, permanent magnet, natural or artificial, has little place in the arts. It cannot be controlled. In common phrase, it cannot be made to "let go" at will. The greatest value of magnetism, as connected with electricity, consists in the fact of the intimate relationship of the two. A magnet may be made at will with the electric current, as described above. A little later we shall see how the process may be reversed, and the magnet be made to produce the most powerful current known, and yet owe its magnetism to the same current.

The word *Magnet* comes from the country of *Magnesia*, where "loadstone" (magnetic iron ore) seems first to have been found. The artificial magnet, as made and used in early experiments and still common as a toy or as a piece in some electrical appliances, is a piece of fine steel, of hard temper, which has been magnetized, usually by having had a current passed through or around it, and sometimes by contact with another magnet. For the singular property of a magnet is that it may continually impart its quality, yet never lose any of its own. Steel alone, of all the metals, has the decided quality of retaining its property of being a magnet. A "bar" magnet is a straight piece of steel magnetized. A "horseshoe" magnet is a bar magnet bent into the form of the letter "U."

Every magnet has two "poles"—the positive, or North pole, and the negative, or South pole. If any magnet, of any size, and having as one piece two poles only, be cut into two, or a hundred pieces, each separate piece will be like the original magnet and have its two poles. The law is arbitrary and invariable under all circumstances, and is a law of nature, as unexplainable and as invariable as any in that mysterious code. All bar magnets, when suspended by their centers, turn their ends to the North and South, a familiar example of this being the ordinary compass. But in magnetism, *like repels like*. The world is a huge magnet. The pole of the magnet which points to the North is not the North pole of the needle as we regard it, but the opposite, the South.

No one can explain precisely why iron, the purer and softer the better, becomes a powerful and effective magnet under the influence of the current, and instantly loses that character when the current ceases, and why steel, the purer and harder the better, at first rejects the influence, and comes slowly under it, but afterwards retains it permanently. Iron and steel are the magnetic metals, but there is a considerable list of metals not magnetic that are better than they as *conductors* of the electric current. In a certain sense they are also the electric metals. A Dynamo, or Motor, made of brass or copper entirely would be impossible. All the phenomena of combined magnetism and electricity, all that goes to make up the field of industrial electric action, would be impossible without the indispensable of ordinary iron, and for the sole reason that it possesses the peculiar qualities, the affinities, described.

* * * * *

There is now an understanding of the electro-magnet, with some idea of the part it may be made to play in the movement of pieces, parts, and machines in which it is an essential. It has been explained how soft iron becomes a magnet, not necessarily by any actual contact with any other magnet, or by touching or rubbing, but by being placed in an electric field. It acquired its magnetism by induction; by *drawing in* (since that is the meaning of the term) the electricity that was around it. But induction has a still wider field, and other characteristics than this alone. Some distinct idea of these may be obtained by supposing a simple case, in which I shall ask the reader to follow me.

[Illustration: DIAGRAM THEORY OF INDUCTION]

Let us imagine a wire to be stretched horizontally for a little space, and its two ends to be attached to the two poles of an ordinary battery so that a current may pass through it. Another wire is stretched beside the first, not touching it, and not connected with any source of electricity. Now, if a current is passed through the first wire a current will also show in the second wire, passing in an *opposite direction* from the first wire's current. But this current in the second wire does not continue. It is a momentary impulse, existing only at the moment of the first passing of the current through the wire attached to the poles of the battery. After this first instantaneous throb there is nothing more. But now cut off the current in the first wire, and the second wire will show another impulse, this time in the *same direction* with the current in the first wire. Then it is all over again, and there is nothing more. The first of these wires and currents, the one attached to the battery poles, is called the *Primary*. The second unattached wire, with its impulses, is called the *Secondary*.

Let us now imagine the primary to be attached to the battery-poles permanently. We will not make or break the circuit, and we can still produce currents, "impulses," in the secondary. Let us imagine the primary to be brought nearer to the secondary, and again moved away from it, the current passing all the time through it. Every time it is moved nearer, an impulse will be generated in the secondary which will be opposite in direction to the current in the primary. Every time it is moved away again, an impulse in the secondary will be in the same direction as the primary current. So long, as before, as the primary wire is quiet, there will be no secondary current at all.

There is still a third effect. If the current in the primary be *increased or diminished* we shall have impulses in the secondary.

This is a supposed case, to render the facts, the laws of induction, clear to the understanding. The experiment might actually be performed if an instrument sufficiently delicate were attached to the terminals of the secondary to make the impulses visible. The following facts are deduced from it in regard to all induced currents. They are the primary laws of induction: —

A current which begins, which approaches, or which increases in strength in the primary, induces, with these movements or conditions, a momentary current in the *opposite direction* in the secondary.

A current which stops, which retires, or which decreases in strength in the primary, induces a momentary current *in the same direction* with the current in the primary.

To make the results of induction effective in practice, we must have great length of wire, and to this end, as in the case of the electro-magnet, we will adopt the spool form. We will suppose two wires, insulated so as to keep them from actually touching, held together side by side, and wound upon a core in several layers. There will then be two wires in the coil, and the opposite ends of one of these wires we will attach to the poles of a battery, and send a current through the coil. This would then be the primary, and the other would be the secondary, as described above. But, since the power and efficiency of an induced current depends upon the length of the secondary wire that is exposed to the influence of the current carried by the primary, we fix two separate coils, one small enough to slip inside of the other. This smaller, inner coil is made with coarser wire than the outer, and the latter has an immense length of finer wire. The current is passed through the smaller, inside coil, and each time that it is stopped, or started, there will be an impulse, and a very strong one, through the outer — the secondary coil. Leave the current uninterrupted, and move the outer coil, or the inner one, back and forth, and the same series of strong impulses will be observed in the coil that has no connection with any source of electricity.

What I have just described as an illustration of the laws governing the production of induced currents, is, in fact, what is known as the *Induction Coil*. In the old times of a quarter of a century ago it was extensively used as an illustrator of the power of the electric current. Sometimes the outer coil contained fifty miles of wire, and the spark, a close imitation of a flash of lightning, would pass between the terminals of the secondary coil held apart for a distance of several feet, and would pierce sheets of plate glass three inches thick. Before the days of practical electric lighting the induction-coil was used for the simultaneous lighting of the gas-jets in public buildings, and is still so used to a limited extent. Its description is introduced here as an illustration of the laws of induction which the reader will find applied hereafter in newer and more effective ways. The commonest instance now of the use of the induction-coil is in the very frequent small machine known as a medical battery. There must be a means of making and breaking the current (the circuit) as described above. This, in the medical battery, is automatic, and it is that which produces the familiar buzzing sound. The mechanism is easily understood upon examination.

* * * * *

At some risk of tediousness with those who have already made an examination of elementary electricity, I have now endeavored to convey to the reader a clear idea of (1), what electricity is, so far as known. (2) Of how the current is conducted, and its influence in the field surrounding the conductor. (3) The nature of the induced current, and the manner in which it is produced. The sum of the information so far may be stated in other words to be how to make an electromagnet, and how to produce an induced current. Such information has an end in view. A knowledge of these two items, an understanding of the details, will be found, collectively or separately, to underlie an understanding of all the machines and appliances of modern electricity, and in all probability, of all those that are yet to come.

But in the prominent field of electric lighting (to which presently we shall come), there is still another principle involved, and this requires some explanation (as well given here as elsewhere) of the current theory as to what electricity is. [Footnote: There are several "schools" among scientists, those who pursue pure science, irrespective of practical applications, and who are rather disposed to narrow the term to include that field alone, that are divided among themselves upon the question of what electricity is. The "Substantialists" believe that it is a kind of matter. Others deny that, and insist that it is a "form of Energy," on which point there can be no serious question. Still others reject both these views. Tesla has said that "nothing stands in the way of our calling electricity 'ether associated with matter, or bound ether.'" Professor Lodge says it is "a form, or rather a mode of manifestation, of the ether" The question is still in dispute whether we have only one electricity or two opposite electricities. The great field of chemistry enters into the discussion as perhaps having the solution of the question within its possibilities. The practical electrician acts upon facts which he knows are true without knowing their cause; empirically; and so far adheres to the molecular hypothesis. The demonstrations and experiments of Tesla so far produce only new theories, or demonstrate the fallacies of the old, but give us nothing absolute. Nevertheless, under his investigations, the possibilities of the near future are widely extended. By means of currents alternating with very high frequency, he has succeeded in passing by induction, through the glass of 1 lamp, energy sufficient to keep a filament in a state of incandescence *without the use of any connecting wires*. He has even lighted a room by producing in it such a condition that an illuminating appliance may be placed anywhere and lighted without being electrically connected with anything. He has produced the required condition by creating in the room a powerful electrostatic field alternating very rapidly. He suspends two sheets of metal, each connected with one of the terminals of the coil. If an exhausted tube is carried anywhere between these sheets, or placed anywhere, it remains always luminous.

Something of the unquestionable possibilities are shown in the following quotation from *Nature*, as expressed in a lecture by Prof. Crookes upon the implied results of Tesla's experiments.

The extent to which this method of illumination may be practically available, experiments alone can decide. In any case, our insight into the possibilities of static electricity has been extended, and the ordinary electric machine will cease to be regarded as a mere toy.

Alternating currents have, at the best, a rather doubtful reputation. But it follows from Tesla's researches that, is the rapidity of the alternation increases, they become not more dangerous but less so. It further appears that a true flame can now be produced without chemical aid—a flame which yields light and heat without the consumption of material and without any chemical process. To this end we require improved methods for producing excessively frequent alternations and enormous potentials. Shall we be able to obtain these by tapping the ether? If so, we may view the prospective exhaustion of our coal-fields with indifference; we shall at once solve the smoke question, and thus dissolve all possible coal rings.

Electricity seems destined to annex the whole field, not merely of optics, but probably also of thermotics.

Rays of light will not pass through a wall, nor, as we know only too well, through a dense fog. But electrical rays of a foot or two wave-length, of which we have spoken, will easily pierce such mediums, which for them will be transparent.

Another tempting field for research, scarcely yet attacked by pioneers, awaits exploration. I allude to the mutual action of electricity and life. No sound man of science indorses the assertion that "electricity is life." nor can we even venture to speak of life as one of the varieties or manifestations of energy. Nevertheless, electricity has an important influence upon vital phenomena, and is in turn set in action by the living being—animal or vegetable. We have electric fishes—one of them the prototype of the torpedo of modern warfare. There is the electric slug which used to be met with in gardens and roads about Hoinsey Rise; there is also an electric centipede. In the study of such facts and such relations the scientific electrician has before him an almost infinite field of inquiry.

The slower vibrations to which I have referred reveal the bewildering possibility of telegraphy without wires, posts, cables, or any of our present costly appliances. It is vain to attempt to picture the marvels of the future. Progress, as Dean Swift observed, may be "too fast for endurance."] As to this, all we may be said to know, as has been remarked, is that it is one of the *forms of energy*, and its manifestations are in the form of *motion* of the minute and invisible atoms of which it is composed. This movement is instantaneously communicated along the length of a conductor. There must, of course, be an end to this process in theory, because all the molecules once moved must return to rest, or to a former condition, before being moved again. Therefore it is necessary to add that when the motion of the last molecule has been absorbed by some apparatus for applying it to utility, the last particles, atoms, molecules, are restored to rest, and may again receive motion from infringing particles, and this transmission of energy along a conductor is continuous—continually absorbed and repeated. This is *dynamic* electricity; not differing in kind, in essence, from any other, but only in application.

If the conductor is entirely insulated, so that no molecular movements can be communicated by it to contiguous bodies, all its particles become energized, and remain so as long as the conductor is attached to a source of electricity. In such a case an additional charge is required only when some of the original charge is taken away, escapes. This is *Static* electricity; the same as the other, but in theory differing in application.

The molecular theory is, unquestionably, tenable under present conditions. It is that to which science has attained in its inquiries to the present date. The electric light is scarcely explainable upon any other hypothesis. The remaining conclusions may be left in abeyance, and without argument.

Science began with static electricity, so called, because its sources were more readily and easily discovered in the course of scientific accidents, as in the original discovery of the property of rubbed amber, etc., and the long course of investigations that were suggested by that antique, accidental discovery. What we know as the dynamic branch of the subject was created by the investigations of Faraday; induction was its mother. It is the practically important branch, but its investigation required the invention of machinery to perform its necessary operations. Between the two branches the sole difference—a difference that may be said not actually to exist—is in *quantity and pressure*.

To the department of static electricity all those industrial appliances first known belong, as the telegraph, electro-plating, etc. I shall first consider this class of appliances and machines. The most important of the class is

[Illustration]

THE ELECTRIC TELEGRAPH.—The word is Greek, meaning, literally, "to write from a distance." But long since, and before Morse's invention, it had come to mean the giving of any information, by any means, from afar. The existence of telegraphs, not electric, is as old as the need of them. The idea of quickness, speedy delivery, is involved. If time is not an object, men may go or send. The means used in telegraphing, in ancient and modern times, have been sound and sight. Anything that can be expressed so as to be read at a distance, and that conveys a meaning, is a telegram. [Footnote: This word is of American coinage, and first appeared in the *Albany Evening Journal*, in 1852. It avoids the use of two words, as "Telegraphic Message," or "Telegraphic Dispatch," and the ungrammatical use of "Telegraph," for a message by telegraph. The new word was at once adopted.] Our plains Indians used columns of smoke, or fires, and are the actual inventors of the *heliograph*, now so called, though formerly meaning the making of a picture by the aid of the sun—photography. The vessels of a squadron at sea have long used telegraphic signals. Some of the celebrated sentences of our history have been written by visual signals, such as "Hold the fort, for I am coming," "Don't give up the ship," etc. Order of showing, positions, and colors are arbitrarily made to mean certain words. The sinking of the "*Victoria*" in 1893, was brought about by the orders conveyed by marine signals. Bells and guns signal by sound. So does the modern electric telegraph, contrary to original design. It is all telegraphy, but it all required an agreed and very limited code, and comparative nearness. None of the means in ancient use were available for the multifarious uses of modern commerce.

As soon as it was known that electricity could be sent long distances over wires, human genius began to contrive a way of using it as a means of conveying definite intelligence. The first idea of the kind was attempted to be put into effect in 1774. This was, however, before the discovery of the electro-magnet (about 1800), or even the Galvanic battery, and it was seriously proposed to have as many wires as there were letters; each wire to have a frictional battery for generating electricity at one end of the circuit, and a pith-ball electroscope at the other. The modern reader may smile at the idea of the hurried sender of a message taking a piece of cat-skin, or his silk handkerchief, and rubbing up the successive letter-balls of glass or sulphur until he had spelled out his telegram. Later a man named Dyer, of New York, invented a system of sending messages by a single wire, and of causing a record to be made at the receiving office by means of a point passing over litmus paper, which the current was to mark by chemical action, the paper passing over a roller or drum during the operation. The battery for this arrangement was also frictional. They knew of no other. Then came the deflected-needle telegraph, first suggested by Ampère, and a few such lines were constructed, and to some extent operated. In one of the original telegraph lines the wires were bound in hemp and laid in pipes on the surface of the ground. The expedient of poles and atmospheric insulation was not thought of until it was adopted as a last resort during the construction of Morse's first line between Washington and Baltimore.

In the year 1832, an American named Samuel F. B. Morse was making a voyage home from Havre to New York in the sailing packet *Sully*. He was an educated man, a graduate of Yale, and an artist, being the holder of a gold medal awarded him for his first work in sculpture, and no want of success drove him to other fields. But during this tedious voyage of the old times in a sailing vessel he seems to have conceived the idea which thenceforth occupied his life. It was the beginning of the present Electric Telegraph. During this same voyage he embodied his notions in some drawings, and they were the beginnings of vicissitudes among the most long-continued and trying for which life affords any opportunity. He abandoned his studies. He paid attention to no other interest. He passed years in silent and lonesome endeavors that seemed to all others useless. He subjected himself to the reproaches of all his friends, lost the confidence of business men, gained the reputation of being a monomaniac, and was finally given over to the following of devices deemed the most useless and unpromising that up to that time had occupied the mind of any man.

The rank and file of humanity had no definite idea of the plan, or of the results that would follow if it were successful. In reality no one cared. It was Morse's enterprise exclusively—a crank's fad alone. There has been no period in the history of society when the public, as a body, was interested in any great change in the systems to which it was accustomed. There is always enmity against an improver. In reality, the question of how much money Morse should make by inventing the electric telegraph was the question of least importance. Yet it was regarded as the only one. He is dead. His profits have gone into the mass, his honors have become international. The patents have long expired. The public, the entire world, are long since the beneficiaries, and the benefits continue to be inconceivably vast. Nothing in all history exceeds in moral importance the invention of the telegraph except the invention of printing with movable types.

[Illustration: AN ELECTRO-MAGNET OF MORSE'S TIME.]

After eight years of waiting, and the repeated instruction of the entire Congress of the United States in the art of telegraphy, that body was finally induced to make an appropriation of thirty thousand dollars to be expended in the construction of an experimental line between Washington and Baltimore. And now begins the actual strangeness of the story of the Telegraph. After many years of toil, Morse still had learned nothing of the efficient construction of an electro-magnet. The magnet which he attempted to use unchanged was after the pattern of the first one ever made—a bent U-shaped bar, around which were a few turns of wire not insulated. The bar was varnished for insulation, and the turns of wire were so few that they did not touch each other. The apparatus would not work at a distance of more than a few feet, and not invariably then. Professor Leonard D. Gale suggested the cause of the difficulty as being in the sparseness of the coils of wire on the magnet and the use of a single-cell battery. He furnished an electro-magnet and battery out of his own belongings, with which the efficiency of the contrivance was greatly increased. The only insulated wire then known was bonnet-wire, used by milliners for shaping the immense flaring bonnets worn by our grandmothers, and when it finally came to constructing the instruments of the first telegraphic system the entire stock of New York was exhausted. The immense stocks of electrical supplies now available for all purposes was then, and for many years afterwards, unknown. Previous to the investigations of Professor Henry, in 1830, only the theory of causing a core of soft iron to become a magnet was known, and the actual magnet, as we make it, had not been made. Morse, in his beginnings, had not money enough to employ a competent mechanic, and was himself possessed of but scant mechanical skill or knowledge of mechanical results. Persistency was the quality by which he succeeded.

[Illustration: DIAGRAM OF MORSE'S INSTRUMENT, 1830, WITH ITS WRITING.]

The battery used first by Morse, as stated, was a single cell. The one made later by his partner, Alfred Vail, the real author of all the workable features of the Morse telegraph, and of every feature which identifies it with the telegraph of the present, was a rectangular wooden box divided into eight compartments, and coated inside with beeswax so that it might resist the action of acids. The telegraphic instrument as made by Morse was a rectangular frame of wood, now in the cabinet of the Western Union Telegraph Company, at New York, which was intended to be clamped to the edge of a table when in use. He knew nothing of the splendid invention since known as the "Morse Alphabet," and the spelling of words in a telegram was not intended by him. His complicated system, as described in his caveat filed by him in 1837, consisted in a system of signs, by which numbers, and consequently words and sentences, were to be indicated. There was then a set of type arranged to regulate and communicate the signs, and rules in which to set this type. There was a means for regulating the movement forward of the rule containing the types. This was a crank to be turned by the hand. The marking or writing apparatus at the receiving instrument was a pendulum arranged to be swung *across* the slip of paper, as it was unwound from the drum, making a zig-zag mark the points of which were to be counted, a certain number of points meaning a certain numeral, which numeral meant a word. A separate type was used to represent each numeral, having a corresponding number of projections or teeth. A telegraphic dictionary was necessary, and one was at great pains prepared by Morse. His process was, therefore, to translate the message to be sent into the numerals corresponding to the words used, to set the types corresponding to those numerals in the rule, and then to pass the rule through the appliance arranged for the purpose in connection with the electric current. The receiver must then translate the message by reference to the telegraphic dictionary, and write out the words for the person to whom the message was sent. This was all changed by Vail, who invented the "dot-and-dash" alphabet, and modified the mechanical action of the instrument necessary for its use. The arrangement of a steel embossing-point working upon a grooved roller—a radical difference—was a portion of this change. The invention of the axial magnet, also Vail's, was another. Morse had regarded a mechanical arrangement for transmitting signals as necessary. Vail, in the practice of the first line, grew accustomed to sending messages by dipping the end of the wire in the mercury cup,—the beginning of the present transmitting instrument, which is also his invention—and Morse's "port-rule," types, and other complicated arrangements, went into the scrap-heap.

[Illustration: MODERN TRANSMITTER.]

Yet there were some strange things still left. The receiving relay weighed 185 pounds. An equally efficient modern one need not weigh more than half a pound. Morse had intended to make a *recording* telegraph distinctively; it was to his mind its chiefest value. Almost in the beginning it ceased to be such, and the recording portion of the instrument has for many years been unknown in a telegraph office, being replaced by the "sounder." This was also the invention of Vail. The more expert of the operators of the first line discovered that it was possible to read the signals *by the sound* made by the armature lever. In vain did the managers prohibit it as unauthorized. The practice was still carried on wherever it could be without detection. Morse was uncompromising in his opposition to the innovation. The wonderful alphabet of the telegraph, the most valuable of the separate inventions that make up the system, was not his conception. The invention of this alphabetical code, based on the elements of time and space, has never met with the appreciation it has deserved. It has been found applicable everywhere. Flashes of light, the raising and lowering of a flag, the tapping of a finger, the long and short blasts of a steam whistle, spell out the words of the English language as readily as does the sounder in a telegraph-office. It may be interpreted by sight, touch, taste, hearing. With a wire, a battery and Vail's alphabet, telegraphy is entirely possible without any other appliances.

[Illustration: MODERN "SOUNDER."]

A brief sketch of the difficulties attending the making of the first practical telegraph line will be interesting as showing how much and how little men knew of practical electricity in 1843. [Footnote: There was no possibility of their knowing more, notwithstanding that, viewed from the present, their inexperienced struggles seem almost pathetic. So, also, do the ideas of Galvani and the experiments and conclusions of all except Franklin, until we come to Faraday. It is one of the features of the time in which we live that, regardless of age, we are all scholars of a new school in which mere diligence and behavior are not rewarded, and in which it is somewhat imperative that we should keep up with our class in an understanding of *what are now the facts of daily life*, wonders though they were in the days of our youth.] To begin with, it was a "metallic circuit;" that is, two wires were to be used instead of one wire and a "ground connection." They knew nothing of this last. Vail discovered and used it before the line was finished. The two wires, insulated, were inclosed in a pipe, lead presumably, and the pipe was placed in the ground. Ezra Cornell, afterwards the founder of Cornell University, had been engaged in the manufacture and sale of a patent plow, and undertook to make a pipe-laying machine for this new telegraph line. After the work had been begun Vail tested and united the conductors as each section was laid. When ten miles were laid the insulation, which had been growing weaker, failed altogether. There was no current. Probably every schoolboy now knows what the trouble was. The earth had stolen the current and absorbed it. The modern boy would simply remark "Induction," and turn his attention to some efficient remedy. Then, there was consternation. Cornell dexterously managed to break the pipe-laying machine, so as to furnish a plausible excuse to the newspapers and such public as there may be said to have been before there was any telegraph line. Days were spent in consultation at the Relay House, and in finding the cause of the difficulty and the remedy. Of the congressional appropriation nearly all had been spent. The interested parties even quarreled, as mere men will under such circumstances, and the want of a little knowledge which is now elementary about electricity came near wrecking forever an enterprise whose vast importance could not be, and was not then, even approximately measured.

[Illustration: ALFRED VAIL.]

Finally, after some weeks delay, it was decided to introduce what has become the most familiar feature of the landscape of civilization, and string the wires on poles. There is little need to follow the enterprise further. Morse stayed with one instrument in the Capitol at Washington, and Vail carried another with him at the end of the line. Already the type-and-rule and all the symbols and dictionaries had been discarded, and the dot-and-dash alphabet was substituted. On April 23d, 1844, Vail substituted the earth for the metallic circuit as an experiment, and that great step both in knowledge and in practice was taken.

Within an incredibly brief space the Morse Electric Telegraph had spread all over the world. No man's triumph was ever more complete. He passed to those riches and honors that must have been to him almost as a fulfilled dream. In Europe his progresses were like those of a monarch. He was made a member of almost all of the learned societies of the world, and on his breast glittered the medals and orders that are the insignia of human greatness. A congress of representatives of ten of the governments of Europe met in Paris in 1858, and it was unanimously decided that the sum of four hundred thousand francs—about a hundred thousand dollars—should be presented to him. He died in New York in 1872.

[Illustration: PROF. HENRY'S ELECTROMAGNET AND ARMATURE]

Yet not a single feature of the invention of Morse, as formulated in his caveat and described in his original patent, is to be found among the essentials of modern telegraphy. They had mostly been abandoned before the first line had been completed, and the arrangements of his associate, Vail, were substituted. Professor Joseph Henry had, in 1832, constructed an electromagnetic telegraph whose signals were made by sound, as all signals now are in the so-called Morse system. He hung a bar-magnet on a pivot in its center as a compass-needle is hung. He wound a U-shaped piece of soft iron with insulated wire, and made it an electro-magnet, and placed the north end of the magnetized bar between the two legs of this electro-magnet. When the latter was made a magnet by the current the end of the bar thus placed was attracted by one leg of the magnet and repelled by the other, and was thus caused to swing in a horizontal plane so that the opposite end of it struck a bell. Thus was an electric telegraph made as an experimental toy, and fulfilling all the conditions of such an one giving the signals by sound, as the modern telegraph does. It lacked one thing—the essential. [Footnote: The details of the construction of the modern telegraph line are not here stated. There are none that change, in principle, the outline above given.]

The Vail telegraphic alphabet had not been thought of. Had such an idea been conceived previously a message could have been read as it is read now, and with the toy of Professor Henry which he abandoned without an idea of its utility or of the possibilities of any telegraph as we have long known them. Morse knew these possibilities. He was one of the innumerable eccentrics who have been right, one of the prophets who have been in the beginning without honor, not only in respect to their own country, but in respect to their times.

[Illustration: DIAGRAM OF TELEGRAPH SYSTEM.]

CHAPTER II.

THE OCEAN CABLE. — The remaining department of Telegraphy is embodied in the startling departure from ancient ideas of the possible which we know as cable telegraphy, the messages by such means being *cablegrams*. About these ocean systems there are many features not applying to lines on land, though they are intended to perform the same functions in the same way, with the same object of conveying intelligence in language, instantly and certainly, but under the sea.

The marine cables are not simple wires. There is in the center a strand of usually seven small copper wires, intended as the conductor of the current. These, twisted loosely into a small cable, are surrounded by repeated layers of gutta-percha, which is, in turn, covered with jute. Outside of all there is an armor of wires, and the entire cable appears much like any other of the wire cables now in common use with elevators, bridges, and for many purposes. In the shallow waters of bays and harbors, where anchors drag and the like occurrences take place, the armor of a submarine cable is sometimes so heavy as to weigh more than twenty tons to the mile.

There are peculiar difficulties encountered in sending messages by an ocean cable, and some of these grow out of the same induction whose laws are indispensable in other cases. The inner copper core sets up induction in the strands of the outer armor, and that again with the surrounding water. There is, again, a species of re-induction affecting the core, so that faint impulses may be received at the terminals that were never sent by the operators. All of these difficulties combined result in what electricians term "retardation." It is one of the departments of telegraphy that, like the unavoidable difficulties in all machines and devices, educates men to their special care, and keeps them thinking. It is one of the natural features of all the mechanical sciences that results in the continual making of improvements.

The first impression in regard to ocean cables would be that very strong currents are used in sending impulses so far. The opposite is true. The receiving instrument is not the noisy "sounder" of the land lines. There was, until recently, a delicate needle which swung to and fro with the impulses, and reflected beams of light which, according to their number and the space between them spelled out the message according to the Vail dot-and-dash alphabet. Now, however, a means still more delicate has been devised, resulting in a faint wavy ink-line on a long, unwinding slip of paper, made by a fountain pen. This strange manuscript may be regarded as the latest system of writing in the world, having no relationship to the art of Cadmus, and requiring an expert and a special education to decipher it. Those faint pulsations, from a hand three thousand miles away across the sea, are the realization of a magic incredible. The necromancy and black art of all antiquity are childish by comparison. They give but faint indications of what they often are—the messages of love and death; the dictations of statesmanship; the heralds of peace or war; the orders for the disposition of millions of dollars.

The story of the laying of the first ocean cable is worthy of the telling in any language, but should be especially interesting to the American boy and girl. It is a story of native enterprise and persistence; perhaps the most remarkable of them all.

The earliest ocean telegraph was that laid by two men named Brett, across the English Channel. For this cable, a pioneer though crossing only a narrow water, the conservative officials of the British government refused a charter. In August, 1850, they laid a single copper wire covered with gutta-percha from Dover in England to the coast of France. The first wire was soon broken, and a second was made consisting of several strands, and this last was soon imitated in various short reaches of water in Europe.

But the Atlantic had always been considered unfathomable. No line had ever sounded its depths, and its strong currents had invariably swept away the heaviest weights before they reached its bed. Its great feature, so far as known, was that strange ocean river first noted and described by Franklin, and known to us as the Gulf Stream. In 1853 a circumstance occurred which again turned the attention of a few men to the question of an Atlantic cable. Lieutenant Berryman, of the Navy, made a survey of the bottom of the Atlantic from Newfoundland to Ireland, and the wonderful discovery was made that the floor of the ocean was a vast plain, not more than two miles below the surface, extending from one continent to the other. This plain is about four hundred miles wide and sixteen hundred long, and there are no currents to disturb the mass of broken shells and unknown fishes that lie on its oozy surface. It was named the "Telegraphic Plateau," with a view to its future use. At either edge of this plateau huge mountains, from four to seven thousand feet high, rise out of the depths. There are precipices of sheer descent down which the cable now hangs. The Azores and Bermudas are peaks of ocean mountains. The warm river known as the Gulf Stream, coming northward meets the ice-bergs and melts them, and deposits the shells, rocks and sand they carry on this plain. When it was discovered the difficulty in the way of an Atlantic cable seemed no longer to exist, and those who had been anxious to engage in the enterprise began to bestir themselves.

Of these the most active was the American, Cyrus W. Field. He began life as a clerk in New York City. When thirty-five years old he became engaged in the building of a land line of telegraph across Newfoundland, the purpose of which was to transmit news brought by a fast line of steamers intended to be established, and the idea is said to have occurred to him of making a line not only so far, but across the sea. In November, 1856, he had succeeded in forming a company, and the entire capital, amounting to 350,000 pounds, was subscribed. The governments of England and the United States promised a subsidy to the stockholders. The cable was made in England. The *Niagara* was assigned by the United States, and the *Agamemnon* by England, each attended by smaller vessels, to lay the cable. In August, 1857, the Niagara left the coast of Ireland, dropping her cable into the sea. Even when it dropped suddenly down the steep escarpment to the great plateau the current still flowed. But through the carelessness of an assistant the cable parted. That was the beginning of mishaps. The task was not to be so easily done, and the enterprise was postponed until the following year.

That next year was still more memorable for triumph and disappointment. It was now designed that the two vessels should meet in mid-ocean, unite the ends of the cable, and sail slowly to opposite shores. There were fearful storms. The huge *Agamemnon*, overloaded with her half of the cable, was almost lost. But finally the spot in the waste and middle of the Atlantic was reached, the sea was still, and the vessels steamed away from each other slowly uncoiling into the sea their two halves of the second cable. It parted again, and the two ships returned to Ireland.

In July they again met in mid-ocean. Europe and America were both charitably deriding the splendid enterprise. All faith was lost. It was known, to journalism especially, that the cable would never be laid and that the enterprise was absurd. But it was like the laying of the first land line. There was a way to do it, existing in the brains and faith of men, though at first that way was not known. From this third meeting the two ships again sailed away, the *Niagara* for America, the *Agamemnon* for Valencia Bay. This time the wire did not part, and on August 29th, 1858, the old world and the new were bound together for the first time, and each could read almost the thoughts of the other. The queen saluted America, and the president replied. There were salutes of cannon and the ringing of bells. But the messages by the cable grew indistinct day by day, and finally ceased. The Atlantic cable had been laid, and—had failed.

Eight years followed, and the cable lay forgotten at the bottom of the sea. The reign of peace on earth and good will to men had so far failed to come and they were years of tumult and bitterness. The Union of the United States was called upon to defend its integrity in a great war. A bitter enmity grew up between us and England. The telegraph, and all its persevering projectors, were almost absolutely forgotten. Electricians declared the project utterly impracticable, and it began, finally, to be denied that any messages had ever crossed the Atlantic at all, and Field and his associates were discredited. It was said that the current could not be made to pass through so long a circuit. New routes were spoken of—across Bering's Strait, and overland by way of Siberia—and measures began to be taken to carry this scheme into effect.

Amid these discouragements, Field and his associates revived their company, made a new cable, and provided everything that science could then suggest to aid final success. This new cable was more perfect than any of the former ones, and there was a mammoth side-wheel steamer known as the *Great Eastern*, unavailable as it proved for the ordinary uses of commerce, and this vessel was large enough to carry the entire cable in her hold. In July, 1865, the huge steamer left Ireland, dropping the endless coil into the sea. The same men were engaged in this last attempt that had failed in all the previous ones. It is one of the most memorable instances of perseverance on record. But on August 6th a flaw occurred, and the cable was being drawn up for repairs. The sound of the wheel suddenly stopped; the cable broke and sunk into the depths. The *Great Eastern* returned unsuccessful to her port.

Field was present on board on this occasion, and had been present on several similar ones. There was, so far as known, no record made by him of his thoughts. There were now five cables in the bed of the Atlantic, and each one had carried down with it a large sum of money, and a still larger sum of hopes. Yet the Great Eastern sailed again in July, 1866, her tanks filled with new cable and Field once more on her decks. It was the last, and the successful attempt. The cable sank steadily and noiselessly into the sea, and on July 26th the steamer sailed into Trinity Bay. The connection was made at Heart's Content, a little New Foundland fishing village, and one for this occasion admirably named. Then the lost cable of 1865 was found, raised and spliced.

In these later times, if a flaw should occur, science would locate it, and go and repair it. Even if this were not true, the fact remains that this last cable, and that of 1865, have been carrying their messages under the sea for nearly thirty years. The lesson is that repeated failures do not mean *final* failure. There is often said to be a malice, a spirit of rebellion, in inanimate things. They refuse to become slaves until they are once and for all utterly subdued, and then they are docile forever. Yet the malice truly lies in the inaptitude and inexperience of men. Had Field and his associates known how to make and lay an Atlantic cable in the beginning as well as they did in the end, the first one laid would have been successful. The years were passed in the invention of machinery for laying, and in improving the construction of each successive cable. Many have been laid since then, certainly and without failure. Men have learned how. [Footnote: At present the total mileage of submarine cables is about 152,000 miles, costing altogether $200,000,000. The length of land wires throughout the world is over 2,000,000 miles, costing $225,000,000. The capital invested in all lines, land and sea, is about $530,000,000.]

Thirteen years were passed in this succession of toils, expenditures, trials and failures. Field crossed the Atlantic more than fifty times in these years, in pursuit of his great idea. At last, like Morse, he was crowned with wealth, success, medals and honors. He was acquainted with all the difficulties. It is now known that he knew through them all that an ocean cable could finally be laid.

THE TELEPHONE.—The telegraph had become old. All nations had become accustomed to its use. More than thirty years had elapsed—a long time in the last half of the nineteenth century—before mankind awoke to a new and startling surprise; the telegraph had been made to transmit not only language, but the human voice in articulate speech. [Footnote: It has been noted that Morse's idea was a *recording* telegraph, that being in his mind its most valuable point, and that this idea has long been obsolete. In like manner, when the Telephone was invented there was a general business opinion that it was perhaps an instrument useful in colleges for demonstrating the wonders of electricity, but not useful for commercial purposes *because it made no record*. "Business will always be done in black and white" was the oracular verdict of prominent and experienced business men. It may be true, but a little conversation across space has been found indispensable. The telephone is a remarkable business success.] The fact first became known in 1873, and was the invention of Alexander G. Bell, of Chicago.

[Illustration: DIAGRAM OF TELEPHONE.—THE BLAKE TRANSMITTER.]

There were several, no one knows how many, attempts to accomplish this remarkable feat previous to the success of Professor Bell. One of these was by Reis, of Frankfort, in 1860. It did not embrace any of the most valuable principles involved in what we know as the telephone, since it could not transmit *speech*. Professor Bell's first operative apparatus was accompanied by simultaneous inventions by Gray, Edison, and others. This remarkable instance of several of the great electricians of the country evolving at nearly the same time the same principal details of a revolutionary invention, has never been fully explained. The first rather crude and ineffective arrangements were rapidly improved by these men, and by others, prominent among whom is Blake, whose remarkable transmitter will be described presently. The best devices of these inventors were finally embodied, and in the resulting instrument we have one of the chiefest of those modern wonders whose first appearance taxed the credulity of mankind. [Footnote: There were, until a recent period, a line of statements, alleged facts and reasonings, that were incredible in proportion to intelligence. The occurrences of recent times have reversed this rule with regard to all things in the domain of applied science. It is the ignorant and narrow only who are incredulous, and the ears of intelligence are open to every sound. All that is not absurd is possible, and all that is possible is sure to be accomplished. The telephone, as a statement, *was* absurd, but not to the men who worked for its accomplishment and finally succeeded. The lines grow narrow. It requires now a high intelligence to decide even upon the fact of absurdity within the domain of natural law.]

In reality the telephone is simple in construction. Workmen who are not accomplished electricians constantly erect, correct and repair the lines and instruments. The machine is not liable to derangement. Any person may use it the first time of trying, and this use is almost universal. Yet it is, from the view of any hour in all the past, an incomprehensible mystery. A moment of reflection drifts the mind backward and renders it almost incredible in the present. The human voice, recognizable, in articulate words, is apparently borne for miles, now even for some hundreds of miles, upon an attenuated wire which hangs silent in the air carrying absolutely nothing more than thousands of little varying impulses of electricity. Not a word that is spoken at one end of it is ever heard at the other, and the conclusion inevitable to the reason of even twenty years ago would be that if one person does not actually hear the other talk there is a miracle. Probably this idea that the voice is actually carried is not very uncommon. The facts seem incomprehensible otherwise, and it is not considered that if that idea were correct it *would* be a miracle.

The entire explanation of the magic of the telephone lies in electrical induction. To the brief explanation of that phenomenon previously given the reader is again referred for a better understanding of what now follows.

But, first, a moment's consideration may be given to the results produced by the use of this appliance, which, as an illustration of the way of the world was an innovation that, had it remained uninvented or impossible, would never have been even desired. One third more business is said now to be transacted in the average day than was possible previously. Since many things can now go on together which previously waited for direction, authority and personal arrangement, a man's business life is lengthened one-third, while his business may mostly be done, to his great convenience, from one place. It has given employment to a large number of persons, a large proportion of whom are young women. The status of woman in the business world has been, fortunately or unfortunately, by so much changed. It has introduced a new necessity, never again to be dispensed with. It has changed the ancient habits, and with them, unconsciously, *the habit of thought.* Contact not personal between man and man has increased. The *thought* of others is quickly arrived at. It has caused us to become more appreciative of the absolute meanings and values of words, without assistance from face, manner or gesture. Laughter may be heard, but tears are unseen. It has induced caution in speech and enforces brevity. While none of its conveniences are now noted, and all that it gives is expected, the telephone, with all its effects, has entered — into the sum of life.

On the wall or table there is a box, and beside this box projects a metal arm. In a fork of this arm hangs a round, black, trumpet-shaped, hard rubber tube. This last is the receiving instrument. It is taken from its arm and held close to the ear. The answers are heard in it as though the person speaking were there concealed in an impish embodiment of himself. Meantime the talking is done into a hole in the side of the box, while the receiver is held to the ear. This is all that appears superficially. An operation incredible has its entire machinery concealed in these simplicities. It is difficult to explain the mystery of the telephone in words — though it has been said to be simple — and it is almost impossible unless the reader comprehends, or will now undertake to comprehend, what has been previously said on the subject of the production of magnetism by a current of electricity, as in the case of the electro-magnet, and on the subject of induction and its laws.

It has been shown that electricity produces magnetism; that the current, properly managed as described, creates instantly a powerful magnet out of a piece of soft iron, and leaves it again a mere piece of iron at the will of the operator. This process also will work backwards. An electric current produces a magnet, and *a magnet also may be made to produce an electric current.* It is one more of the innumerable, almost universal, cases where scientific and mechanical processes may be reversed. When the dynamo is examined this process is still further exemplified, and when we examine the dynamo and the motor together we have a striking example of the two processes going on together.

The application of this making of a current, or changing its intensity, in the telephone, is apparently totally unlike the continuous manufacture of the induced current for daily use by means of the steam engine and dynamo. But it is in exact accord with the same laws. It will, perhaps, be more readily understood by recalling the results of the experiment of the two wires, where it was found that an *approach to*, or a *receding from*, a wire carrying a current, produces an impulse over the wire that has by itself no current at all. Now, it must be added to that explanation that if the battery were detached from that conducting wire, and if, instead of its being a wire for the carrying of a battery current *it were itself a permanent magnet*, the same results would happen in the other wire if it were rapidly moved toward and away from this permanent magnet. If the reader should stretch a wire tightly between two pegs on a table, and should then hold the arms of a common horseshoe magnet very near it, and should twang the stretched wire with his finger, as he would a guitar string, the electrometer would show an induced alternate current in the wire. Since this is an illustration of the principle of the dynamo, stated in its simplest form, it may be well to remember that in this manner—with the means multiplied and in all respects made the most of—a very strong current of electricity may be evolved without any battery or other source of electricity except a magnet. In connection with this substitution of a magnet for a current-carrying wire, it must be remembered that moving the magnet toward or from the wire has the same result as moving the wire instead. It does not matter which piece is moved.

In addition to the above, it should be stated that not only will an induced current be set up in the wire, but also *the magnetism in the magnet will be increased or diminished as the tremblings of the wire cause it to approach or recede from it*. Therefore if a wire be led away from each pole of a permanent magnet, and the ends united to form a circuit, an induced current will appear in this wire if a piece of soft iron is passed quickly near the magnet.

There is an essential part of the telephone that it is necessary to go outside of the field of electricity to describe. It is undoubtedly understood by the reader that all sound is produced by vibrations, or rapid undulations, of the surrounding air. If a membrane of any kind is stretched across a hoop, and one talks against it, so to speak, the diaphragm or membrane will be shaken, will vibrate, with the movement of the air produced by the voice. If a cannon be fired all the windows rattle, and are often broken. A peal of thunder will cause the same jar and rattle of window panes, manifestly by what we call "sound"—vibrations of the air. The window frame is a "diaphragm." The ear is constructed on the same principle, its diaphragm being actually moved by the vibrations of air, being what we call hearing. With these facts about sound understood in connection with those given in connection with the substitution of a magnet for a battery current, it is entirely possible for any non-expert to understand the theory of the construction of the telephone.

In the Bell telephone, now used with the Blake transmitter [which differs somewhat from the arrangement I shall now describe] a bar magnet has a portion of its length wound with very fine insulated wire. Across the opposite end of this polarized [Footnote: "Polarized" means magnetized; having the two poles of a permanent magnet. The term is frequently used in descriptions of electrical appliances. Instead of using the terms *positive* and *negative*, it is also customary to speak of the "North" or the "South" of a magnet, battery or circuit.] magnet, crosswise to it, and very close, there is placed a diaphragm of thin sheet iron. This is held only around its edge, and its center is free to vibrate toward and from the end of this polarized magnet. This thin disc of iron, therefore, follows the movements, the "soundwaves," of the air against it, which are caused by the human voice. We have an instance of apiece of soft iron moving toward, and away from, a magnet. It moves with a rapidity and violence precisely proportioned to the tones and inflections of the voice. Those movements are almost microscopic, not perceptible to the eye, but sufficient.

The approaching and receding have made a difference, in the quality of the magnet. Its magnetism has been increased and diminished, and the little coil of insulated wire around it has felt these changes, and carried them as impulses over the circuit of which it is a part. In that circuit, at the other end, there is a precisely similar little insulated coil, upon a precisely similar polarized magnet. These impulses pass through this second coil, and increase or diminish the magnetism in the magnet round which it is coiled. That, in turn, affects by magnetic attraction the diaphragm that is arranged in relation to its magnet precisely as described for the first. The first being controlled as to the extent and rapidity of its movements by the loudness and other modifications of the voice, the impulses sent over the circuit vary accordingly. As a consequence, so does the strength of the magnet whose coil is also in the circuit. So, therefore, does its power of attraction over its diaphragm vary. The result is that the movements that are caused in the first diaphragm by the voice, are caused in the second by an *attraction* that varies in strength in proportion to the vibrations of the voice speaking against the first diaphragm.

This is the theory of the telephone. The sounds are not carried, but *mechanically produced* again by the rattle of a thin piece of iron close to the listener's ear. The voice is full, audible, distinct, as we hear it naturally, and as it impinges upon the transmitting diaphragm. In reproduction at the receiving instrument it is small in volume; almost microscopic, if the phrase may be applied to sound. We hear it only by placing the ear close to the diaphragm. It will be seen that this is necessarily so. No attempts to remedy the difficulty have so far been successful. There is no means of reproducing the volume of the voice with the minute vibrations of a little iron disc.

In actual service an electro-magnet is used instead of, or in addition to, the bar magnets described above. A steady flow from a battery is passed through an instrument which throws this current into proper vibrations by stopping the flow of the current at each interval between impulses. There is a piece of carbon between the diaphragm and its support. The wires are connected with the diaphragm and its support, and the current passes through the carbon. When the diaphragm vibrates, the carbon is slightly compressed by it. Pressure reduces its resistance, and a greater current passes through it and over the wires of the circuit for the instant during which the touch remains. This is the Blake transmitter. It should be explained that carbon stands low on the list of conductors of electricity. The more dense it is, the better conductor. The varying pressures of the diaphragm serve to produce this varying density and the consequent varying impulses of the current which effect the receiving diaphragm.

The transmitter, as above described, is in the square box, and its round black diaphragm may be seen behind the round hole into which one talks. [Footnote: Shouting into a telephone doubtless comes of the idea, unconscious, that one is speaking to a person at a distance. To speak distinctly is better, and in an ordinary tone.] The receiver is the trumpet-shaped tube which hangs on its side, and is taken from its hook to be used. The call-bell has nothing to do with the telephone. It is operated by a small magneto-generator,—a very near relative of the dynamo-the current from which is sent over the telephone circuit (the same wires) when the small crank is turned. Sometimes the question occurs: "Why ring one's own bell when one desires to ring only that at the central office?" The answer is that both bells are in the same circuit. If the circuit is uninterrupted your bell will ring when you ring the other, and a bell at each end of your circuit is necessary in any case, else you could not yourself be called.

When the receiving instrument is on its hook its weight depresses the lever slightly. This slight movement *connects* the bell circuit and *disconnects* the telephone circuit. Take it off the hook and the reverse is effected.

The long-distance telephone differs from the ordinary only in larger conductors, improved instruments, and a metallic circuit—

two wires instead of the ordinary single wire and ground connections.

[Illustration: TELEAUTOGRAPH TRANSMITTING INSTRUMENT.]

THE TELAUTOGRAPH.—This, the latest of modern miracles in the field of electricity, comes naturally after the telegraph and telephone, since it supplements them as a means of communication between individuals. It also is the invention of Prof. Elisha Gray, who seems to be as well the author of the name of his extraordinary achievement. It is not the first instrument of the kind attempted. The desire to find a means of writing at a distance is old. Bain, of Edinburgh, made a machine partially successful fifty years ago. Like the telegraph as intended by Morse, there was the interposition of typesetting before a message could be sent. It did not write, or follow the hand of the operator in writing, though it did reproduce at the other end of the circuit in facsimile the faces of the types that had been set by the sender. It was a process by electrolysis, well understood by all electricians. Several of this variety of writing telegraphs have been made, some of them almost successful, but all lacking the vital essential. [Footnote: The lack of *one vital essential* has been fatal to hundreds of inventions. Inventors unconsciously follow paths made by predecessors. The entire class of transmitting instruments must dispense with tedious preliminaries, and must use *words*. Vail accomplished this in telegraphy. Bell and others in the telephone, and Gray has borne the same fact in mind in the present development of the telautograph.] In 1856 Casselli, of Florence, made a writing telegraph which had a pendulum arrangement weighing fourteen pounds. Only one was ever made, but it resulted in many new ideas all pertaining to the facsimile systems—the following of the faces of types—and all were finally abandoned.

The invention of Gray is a departure. The sender of a message sits down at a small desk and takes up a pencil, writing with it on ordinary paper and in his usual manner. A pen at the other end of the circuit follows every movement of his hand. The result is an autograph letter a hundred miles or more away. A man in Chicago may write and sign a check payable in Indianapolis. Personal directions may be given authoritatively and privately. As in the case of the telephone, no intervening operator is necessary. No expertness is required. Even the use of the alphabet is not necessary. A drawing of any description, anything that can be traced with a pen or pencil, is copied precisely by the pen at the receiving desk. The possibilities of this instrument, the uses it may develop, are almost inconceivable. It might be imagined that the lines drawn would be continuous. On the contrary, when the pen is lifted by the writer at the sending desk it also lifts itself from the paper at that of the receiver.

The action of the telautograph depends upon the variations in magnetic strength between two small electro-magnets. It has been seen that an electro-magnet exerts its attractive force in proportion to the current which passes through its coil. To use a phrase entirely non-technical, it will "pull" hard or easy in proportion to the strength of the passing current. This fact has been observed as the cause of action in the telephone, where one diaphragm, moved by the air-vibrations caused by the voice, causes a varying current to pass over the wire, attracting the other diaphragm less or more as the first is moved toward or away from its magnet. In the telautograph the varying currents are caused not by the diaphragm influenced by the voice, but *by a pencil moved by the hand*.

To show how these movements may be caused let us imagine a case that may occur in nature. It is an interesting mechanical study. There is an upright rush or reed growing in the middle of a running stream. The stem of this rush has elasticity naturally; it has a tendency to stand upright; but it bends when there is a current against it. It is easy enough to imagine it bending down stream more or less as the current is more or less strong.

Imagine now another stream entering the first at right angles to it, and that the rush stands in the center of both currents. It will then bend to the force of the second stream also, and the direction in which it will lean will be a compromise between the forces of the two. Lessen the flow of the current in one of the streams, and the rush will bend a little less before that current and swing around to the side from which it receives less pressure. Cut off either of the currents entirely, and it will bend in the direction of the other current only. In a word, *if the quantity or strength of the current of both streams can be controlled at will, the rush can be made to swing in any direction between the two, and its tip will describe any figure desired, aided, of course, by its own disposition to stand upright when there is no pressure.*

Let us imagine the rush to be a pen or pencil, and the two streams of water to be two currents of electricity having power to sway and move this pencil in proportion to their relative strength, as the streams did the rush. Imagine further that these two currents are varied and changed with reference to each other by the movements of a pen in a man's hand at another place. It is an essential part of the mechanism of the telautograph, and the movement is known among mechanicians as "compounding a point."

Gray, while using the principles involved in compounding a point, seems to have discarded the ways of transmitting magnetic impulses of varying strength commonly in use. His method he calls the "step-by-step" principle, and it is a striking example of what patience and ingenuity may accomplish in the management of what is reputedly the most elusive and difficult of the powers of nature. The machine was some six years in being brought into practical form, and was perfected only after a long series of experiments. In its operation it deals with infinitesimal measurements and quantities. The first attempts were on the "variable current" system, which was later discarded for the "step-by-step" plan mentioned.

In writing an ordinary lead pencil may be used. From the point of this two silk cords are extended diagonally, their directions being at right angles to each other, and the ends of these cords enter openings made for them in the cast iron case of the instrument on each side of the small desk on which the writing is done.

Inside the case each cord is wound on a small drum which is mounted on a vertical shaft. Now if the pencil-point is moved straight upward or downward it is manifest that both shafts will move alike. If the movement is oblique in any direction, one of the shafts will turn more than the other, and the degree of all these turnings of each shaft in reference to the other will be precisely governed by the direction in which the pencil-point is moved.

[Illustration: DIAGRAM OF MECHANICAL TELAUTOGRAPH. BOW-DRILL ARRANGEMENT.]

Now, suppose each shaft to carry a small, toothed wheel, and that upon these teeth a small arm rests. As the wheel turns this arm will move as a pawl does on a ratchet. Imagine that at each slight depression between the ratchet-teeth it breaks a contact and cuts off a current, and at each slight rise renews the contact and permits a current to pass. This current affects an electro-magnet—one for each shaft—at the receiving end, and each of these magnets, when the current is on, attracts an armature bearing a pawl, which, being lifted, allows the notched wheel, upon which it bears, to turn *to the extent of one notch*. The arrangement may be called an electric clutch, that may be arranged in many ways, and the detail of its action is unimportant in description, so that it be borne in mind that *each time a notch is passed in turning the shaft by drawing upon or relaxing the cords attached to the pencil-point*, an impulse of electricity is sent to an electro-magnet and armature which allows *a corresponding wheel and its shaft to turn one notch, or as many notches, as are passed at the transmitting shaft*. In moving the pencil one inch to one side, we will suppose it permits the shaft on which the cord is wound to turn forty notches. Then forty impulses of electricity have been sent over the wire, the clutch has been released forty times, and the shaft to which it is attached has turned precisely as much as the shaft has which was turned, or was allowed to turn, by the cord wound upon it and attached to the pencil.

It will be remembered that the arrangement is double. There are two shafts operated by the writer's pencil—one on each side of it. Two corresponding shafts occupy relative positions in respect to the automatic pen of the receiving instrument. There are two circuits, and two wires are at present necessary for the operation of the instrument. It remains to describe the manner of operating the automatic pen by connection with its two shafts which are turned by the step-by-step arrangement described, precisely as much and at the same time as those of the transmitting instrument are.

[Illustration: WORK OF THE TELAUTOGRAPH. COLUMBIAN EXPOSITION, 1893.]

To each shaft of the receiving instrument is attached an aluminum pen-arm by means of cords, each arm being fixed, in regard to its shaft, as a bow drill is in regard to its drill. These arms meet in the center of the writing tablet, V-shaped, as the cords are with relation to the writer's pencil in the sending instrument. A small tube conveys ink from a reservoir along one of the pen-arms, and into a glass tube upright at the junction of the arms. This tube is the pen. Now, let us imagine the pencil of the writer pushed straight upward from the apex of the V-shaped figure the cords and pencil-point make on the writing desk. Then both the shafts at the points of the arms of the V will rotate equally. [Footnote: See diagram of mechanical Telautograph, and of bow drill. In the latter, in ordinary use, the stick and string; rotate the spool. Rotating the spool will, in turn, move the stick and string, and this is its action in the pen-arms of the Telautograph.] The number of impulses sent from each of these shafts, by the means explained, will be equal. Each of the shafts of the receiving instrument will rotate alike, and each draw up its arm of the automatic pen precisely as though one took hold of the points of the two legs of the V, and drew them apart to right and left in a straight line. This moves the apex of the V, with its pen, in a straight line upward at the same time the writer at the sending instrument pushed his pencil upward. If this one movement, considered alone, is understood, all the rest follow by the same means. This is, as nearly as it may be described without the use of technical mechanical terms, the principle of the telautograph. It must be seen that all that is necessary to describe any movement of the sender's pencil upon the paper under the receiving pen is that the rotating upright shafts of the latter should move precisely as much, and at the same time, with those two which get their movement from the wound cords and attached pencil-points in the hand of the writer.

Only one essential item of the movement remains. The shafts of both instruments must be rotated by some separate mechanical agency, capable of being automatically reversed. By an arrangement unnecessary to explain in detail, the pencil of the writer lifted from the paper resting on the metallic table which forms the desk; results in the automatic lifting of the pen from the paper at the receiving desk.

* * * * *

Prof. Elisha Gray was born in 1835, in Ohio. He was a blacksmith, and later, a carpenter. But he was given to chemical and mechanical experiments rather than to the industries. When twenty-one, he entered Oberlin College, remaining there five years, and earning all the money he spent. He devoted his time chiefly to studies of the physical sciences. As a young man he was an invalid. Later he was not remarkably successful in business, failing several times in his beginnings. His first invention was a telegraph self-adjusting relay. It was not practically successful. Afterwards he was employed with an electrical manufacturing company at Cleveland and Chicago. Most of his earlier inventions in the line of electrical utility are not distinctively known. He has never been idle, and they all possessed practical merit. For many years before he was known as the wizard of the telautograph, he was foremost in the ranks of physicists and electricians. He is not a discoverer of great principles, but is professionally skillful and accomplished, and eminently practical. His every effort is exerted to avoid intricacy and clumsiness in machinery. In 1878 he was awarded the grand prize at the Paris Exposition, and was given the degree of Chevalier and the decorations of the Legion of Honor by the French Government, and again in 1881, at the Electrical Exposition at Paris, he was honored with the gold medal for his inventions. He secured the degree of A.M. at Oberlin College, and was the recipient of the degree of Ph.D. from the Ripon (Wis.) College. For years he was connected with those institutions as non-resident Lecturer in Physics. Another University gave him the degree of LL.D. He is a member of the American Philosophical Society, the Society of Electrical Engineers of England, and the Society of Telegraph Engineers of London. He received an award and a certificate from the Centennial Exposition for his inventions in electricity.

The same lesson is to be gathered from his career, so far, that is given by the life of every noted American. It means that money, family, prestige, have no place as leverages of success in any field. The rule is toward the opposite. The qualities and capacities that win do so without these early advantages, and all the more surely because there is an inducement to use them. There is no "luck."

CHAPTER III.

THE ELECTRIC LIGHT.

[Illustration]

It has been stated that modern theory recognizes two classes of electricity, the *Static* and the *Dynamic*. The difference is, however, solely noticeable in operation. Of the dynamic class there can be no more common and striking example than the now almost universal electric light. Yet, with a sufficient expenditure of chemicals and electrodes, and a sufficient number of cells, electric lighting, either arc or incandescent, can be as effectively accomplished as with the current evolved by a powerful dynamo. [Footnote: As an illustration of the day of beginnings, a few years ago the *thalus*, or lantern, the pride of the rural Congressman, on the dome of the Capitol at Washington was lighted by electricity, and an immense circular chamber beneath the dome was occupied by hundreds of cells of the ordinary form of battery. The lamps were of the incandescent variety, and what we now know as the filament was platinum wire. Vacuum bulb, filament, carbon, dynamo, were all unknown. But the current, and the heat of resistance, and every fact now in use in electric lighting, were there in operation.]

The reader will understand that modern dynamic electricity owes its development to the principle of economy in production. Practical science most effectively awakens from its lethargy at the call of commerce. Nevertheless, from the earliest moment in which it became known that electricity was akin to heat—that an interruption of the easy passage of a current produced heat—the minds of men were busy with the question of how to turn the tremendous fact to everyday use. Progress was slow, and part of it was accidental. The great servant of modern mankind was first an untrained one. It was a marked advance when the gaslights in a theater could be all lighted at once by means of batteries and the spark of an induction coil. The bottom of Hell Gate, in New York harbor, was blown out by

Gen. Newton by the same means, and would have been impossible otherwise. But these were only incidents and suggestions. The question was how to make this instantaneous spark *continuous*. There was pondering upon the fact that the only difference between heat and electricity is one of molecular arrangement. Heat is a molecular motion like that of electricity, without the symmetry and harmony of action electricity has. The vibrations of electricity are accomplished rapidly, and without loss. Those of heat are slow, and greatly radiated. *When a current of electricity reaches a place in the conductor where it cannot pass easily, and the orderly vibrations of its molecules are disturbed, they are thrown into the disorderly motion known as heat.* So, when the conductor is not so good; when a large wire is reduced suddenly to a small one; when a good conductor, such as copper, has a section of resisting conduction, such as carbon; heat and light are at once evolved at that point, and there is produced what we know as the electric light. However concealed by machinery and devices, and all the arrangements by which it is made more lasting, steady, economical and automatic, it is no more nor less than this. *The difference between heat and electricity is only a difference in the rates of vibration of their molecules.* Whatever the theory as to molecules, or essence, or actual nature and origin, the practical fact that heat and light are the results of the circumstances described above remains. This has long been known, and the question remained how to produce an adequate current economically. The result was the machine we know as the Dynamo.

The first electric light was very brief and brilliant and was made by accident. Sir Humphrey Davy, in 1809, in pulling apart the two ends of wires attached to a battery of two thousand small cells, the most powerful generator that had been made to that time, produced a brief and brilliant spark, the result of momentarily *imperfect contact*. Every such spark, produced since then innumerable times by accident, is an example of electric lighting. There are now in use in the United States some two million arc lights and nearly double that number of incandescent.

There are two principal systems of electric lighting; one is by actually burning away the ends of carbon-points in the open air. This is the "arc." The other is by heating to a white heat a filament of carbon, or some substance of high resistance, in a glass bulb from which the air has been exhausted. This is the "incandescent."

[Illustration: THE INCANDESCENT LIGHT]

In the arc light the current passes across an *imperfect contact*, and this imperfection consists in a gap of about one-sixteenth of an inch between the extremities of two rods of carbon carrying a current. This small gap is a place of bad conduction and of the piling up of atoms, producing heat, burning, light. In the body of the lamp there are appliances for the automatic holding apart of the two points of the carbon, and the causing of them to continually creep together, yet never touch. Many devices have been contrived to this end. With all theories and reasons well known, and all effects accurately calculated, upon this small arrangement depends the practical utility of the arc light. The best arrangement is the invention of Edison, and is controlled most ingeniously by the current itself, acting through the increased difficulty of its passage when the two carbon-points are too far apart, and the increased ease with which it flows when they are too near together. The current, in leaping the small gap between the carbon-points, takes a *curved* path, hence the name "arc" light. In passing from the positive to the negative carbon it carries small particles of incandescent carbon with it, and consequently the end of the positive carbon is hollowed out, while the end of the negative is built up to a point.

The incandescent light is in principle the same as the arc, produced by the same means and based upon the same principle of impediment to the free passage of the current. It was first produced by heating with the current to incandescence a fine platinum wire. As stated above, electricity that quietly traverses a large wire will suddenly develop great heat upon reaching a point where it is called upon to traverse, a smaller one. Platinum was attempted for this place of greater resistance because of its qualities. It does not rust, has a low specific heat, and is therefore raised to a higher temperature with less heat imparted. But it was a scarce and expensive material, and so long as it was heated to incandescence in the open air, that is, so long as its heat was fed as other heat is, by oxygen, it was slowly consumed. Platinum is no longer in the field of electric lighting, and the substitute which takes its place in the present incandescent lamp, and which is known as a "filament," is not heated in contact with the air. The experiments and endeavors that brought this result constitute the story of the incandescent lamp.

The result is due to the patient intelligence of the American scientist and inventor, Thomas A. Edison. After all the absolute essentials of a practical incandescent lamp had been thought out; after the qualities and characteristics of the current were all known under the circumstances necessary to its use in lighting, the practical accomplishment still remained. Edison is said to have once worked for several weeks in the making of a single loop-shaped carbon filament that would bear the most delicate handling. This was then carefully carried to a glass-worker to be inclosed in a bulb, and at the first movement he broke it, and the work must be done over and done better. It finally was. The little pear-shaped bulb with its delicate loop of filament, which cost months of toil and experiment at first, is now a common article, manufactured at an absurdly small cost, packed in barrelfuls and shipped everywhere, and consumed by the million. A means has been found for producing the vacuum of its interior rapidly, cheaply and thoroughly, and the beautiful incandescent glow hangs in lines and clusters over the civilized world. The phenomenon of incandescence without oxygen seems peculiar to these lights alone. [Footnote: The "electric field," previously explained, seemed to exist by giving a magnetic quality to the surrounding air. It would be as true if one should speak of a magnetized vacuum, since the same field would exist in that as in surrounding air.]

So simple are great facts when finally accomplished that there remains little to add on the subject of the mechanism of the electric light. The two varieties, arc and incandescent, are used together as most convenient, the large and very brilliant arc being especially adapted to out-of-doors situations, and the gentler, steadier and more permanent glow of the incandescent to interiors. The latter is also capable of a modification not applicable to the arc. It can, in theaters and other buildings, be "turned down" to a gentle, blood-red glow. The means by which this is accomplished is ingenious and surprising, since it means that the supply of electricity over a wire—seemingly the most subtle and elusive essence on earth—may be controlled like a stream from a cock, or the gas out of a burner. But this reduction of the current that makes the red glow in the clusters in a theater is by no means the only instance. The trolley-car, and even the common motor, may be made to start very slowly, and the unseen current whose touch kills is fed to its consumer at will.

[Illustration]

THE DYNAMO.—To the man who has been all his life thinking of the steam engine as the highest and almost only embodiment of controlled mechanical power, another machine, both supplementary to the steam engine and far excelling it, whose familiar *burring* sound is now heard in almost every village in the United States and has become the characteristic sound of modern civilization, must constitute a source of continual question and surprise. To be accustomed to the dynamo, to look upon it as a matter of course and a conceded fact, one must have come to years of maturity and found it here.

Its practical existence dates back at furthest to 1870. Yet it is based upon principles long since known, and can scarcely be said to be the invention of any one mind or man. Its lineal ancestor was the *magneto-electric machine*, in the early construction of which figure the names of Siemens, Wilde, Ladd, and earlier and later electricians. Kidder's medical battery used forty years ago or more, and still used and purchasable in its first form, was a dynamo. A footnote in a current encyclopedia states that: "An account of the Magneto-electric machine of M. Gramme, in the London *Standard* of April 9th, 1873, confirmed by other information, leads to the belief that a decided improvement has been made in these machines." The word "dynamo" was then unknown. Later, Edison, Weston, Thompson, Hopkinson, Ferranti and others appear as improvers in the mechanism necessary for best developing a well-known principle, and many of these improvements may be classed among original inventions. As soon as the magneto-electric machine attained a size in the hands of experimenters that took it out of the field of scientific toys it began to be what we now know as a dynamo. A paragraph in the encyclopedia referred to says, in speaking of Ladd, of London, "These developments of electric action are not obtained without corresponding expenditure of force. The armatures are powerfully attracted by the magnets, and must be forcibly pulled away. Indeed, one of Wilde's machines, when producing a very intense electric light, required about five horse power to drive it."

[Illustration: MAGNETO-ELECTRIC MACHINE. THE PREDECESSOR OF THE DYNAMO.]

Thus was the secret in regard to electric power unconsciously divulged some twenty years ago.

In all nature there is no recipe for getting something for nothing. The modern dynamo, apparently creating something out of nothing, like all other machines *gives back only what is given to it*, minus a fair percentage for waste, loss, friction, and common wear. Its advantages amount to a miracle of convenience only. So far as power is concerned, it merely transfers it for long distances over a single wire. So far as light is considered, it practically creates it where wanted, in new and convenient forms, with a new intensity and beauty, but with the same expenditure of transmitted energy in the form of burned coal as would be used in manufacturing the gas that was new, wonderful, and a luxury at the beginning of the century.

The dynamo is the most prominent instance of actual mechanical utility in the field of electrical induction. It seems almost incredible that the apparently small facts discovered by Faraday, the bookbinder, the employé of Sir Humphrey Davy at weekly wages the struggling experimenter in the subtleties of an infant giant, should have produced such results within sixty years. [Footnote: Faraday was not entirely alone in his life of physical research. He was associated with Davy, and quarreled with him about the liquefaction of chlorine and other gases, and was the companion of Wallaston, Herschel, Brand, and others. In connection with Stodart, he experimented with steel, with results still considered valuable. The scientific world still speaks of his quarrel with Davy with regret, since the personalities of great men should be free from ordinary weaknesses. But Lady Davy was not a scientist, and while the brilliant young mechanic was in her husband's employment for scientific purposes she insisted upon treating him as a servant, whereat the independence of thinking which made him capable of wandering in fields unknown to conventionality and routine blazed into natural resentment. The quarrel of 1823 must have been greatly augmented, in the lady's eyes, in 1824, for in that year Faraday was made a member of the Royal Society.

In his lectures and public experiments he was greatly assisted by a man now almost forgotten, an "intelligent artilleryman" named Andersen. This unknown soldier with a taste for natural science doubtless had his reward in the exquisite pleasure always derived from the personal verification of facts hitherto unknown. There is often a pecuniary reward for the servant of science. Just as often there is not, and the work done has been the same.

It was on Christmas morning, 1821, that Faraday first succeeded in making a magnetic needle rotate around a wire carrying an electric current. He was the discoverer of benzole, the basis of our modern brilliant aniline dyes. In 1831 he made the discovery he had been leading to for many years—that of magneto-electric induction. All we have of electricity that is now a part of our daily life is the result of this discovery.

Faraday was born in 1791, and died August, 1867, in a house presented to him by Victoria, who had not the same opinion of his relations to the aristocracy that Lady Davy seems to have had. His insight into science was something explainable only on the supposition that he was gifted with a kind of instinct. He was a scientific prophet. A man who could, in 1838, foresee the ocean cable, and describe those minute difficulties in its working that all in time came true, must be classed as one of the great, clear, intuitive intellects of his race. He was in youth apprenticed to a bookbinder, "and many of the books he bound he read." A line in his indentures says: "In consideration of his faithful service, no premium is to be given." When these words were written there was no dream that the "faithful service" should be for all posterity.]

[Illustration: Faraday's Spark. Striking the leg of a horseshoe magnet with an iron bar wound with insulated wire causes a contact between loose end of wire and small disc, and a spark.

Faraday's First Magneto-Electric Experiment. A horseshoe magnet passed near a bent soft iron wound with insulated wire caused an induced current in the wire.

TWO OF FARADAY'S EARLY EXPERIMENTS IN INDUCTION.]

He who made the first actual machine to evolve a current in compliance with Faraday's formulated laws was an Italian named Pixü, in 1832. His machine consisted of a horseshoe magnet set on a shaft, and made to revolve in front of two cores of, soft iron wound with wire, and having their ends opposite the legs of the magnet. Shortly after Pixü, the inventors of the times ceased to turn the magnet on a shaft, and turned the iron cores instead, because they were lighter. In like manner, the huge field magnets of a modern dynamo are not whirled round a stationary armature, but the armature is whirled within the legs of the magnet with very great rapidity. The next step was to increase the number of magnets and the number of wire-wound iron cores—bobbins. The magnets were made compound, laminated; a large number of thin horseshoe magnets were laid together, with opposite poles touching. These were all comparatively small machines—what we now, with some reason, regard as having been toys whose present results were rather long in coming.

[Illustration: THE SIEMENS' ARMATURE AND WINDING. THE FIRST STEP TOWARD
THE MODERN DYNAMO.]

Then came Siemens, of Berlin, in 1857. He was probably the first to wind the iron core, what we now call the *armature*, with wire from end to end, *lengthwise*, instead of round and round as a spool. This resulted, of course, in the shaft of the armature being also placed crosswise to the legs of the magnet, as it is in the modern dynamo. One of the ends of the wire used in this winding was fastened to the axle of the armature, and the other to a ring insulated from the shaft, but turning with it. Two springs, one bearing on the shaft and the other on the ring, carried away the current through wires attached to them. Siemens also originated the mechanical idea of hollowing out the legs of the magnet on the inside for the armature to turn in close to the magnet, almost fitting. It was the first time any of these things had been done, and their author probably had no idea that they would be prominent features of the dynamo of a little later time, in all essentials closely imitated.

[Illustration: DIAGRAM OF SHAFT, SPLIT RING AND "BRUSHES."]

It will be guessed from what has been previously said on the subject of induction that the currents from such an electro-magnetic machine would be alternating currents, the impulses succeeding each other in alternate directions. To remedy this and cause the currents to flow always in the same direction, the "*commutator*" was devised. The ring mentioned above was split, and the two springs both bore upon it, one on each side. The ends of the wires were both fastened to this ring. The springs came to be known as "brushes." The effect was that one of them was in the insulated space between the split halves of the ring while the other was bearing on the metal to which the wire was attached. This action was alternate, and so arranged that the current carried away was always direct. When an armature has a winding of more than one wire, as the practical dynamo always has, the insulated ring is divided into as many pieces as there are wires, and the two brushes act as above for the entire series.

Pacinotti, of Florence, constructed a magneto-electric machine in which the current flows always in one direction without a commutator. It has what is known as a *ring armature*, and is the mother of all dynamos built upon that principle. It is exceedingly ingenious in construction, and for certain purposes in the arts is extensively used. A description of it is too technical to interest others than those personally interested in the class of dynamo it represents.

Wilde, of Manchester, England, improved the Siemens machine in 1866 by doing that which is the feature that makes possible the huge "field magnet" of the modern dynamo, which is not a magnet at all, strictly speaking. He caused the current, after it had been rectified by the commutator, to return again into coils of wire round the legs of his field magnets, as shown in the diagram. This induced in them a new supply of magnetism, and this of course intensified the current from the armature. It is true he had a separate smaller magneto-electric machine, with which he evolved a current for the coil around the legs of the field magnet of a greatly larger machine upon which he depended for his actual current, and that he did not know, although he was practically doing the same thing, that if he should divert this current made by the larger machine itself back through the coils of its field magnet, he would not need the extra small machine at all, and would have a much more powerful current.

[Illustration: SIMPLEST FORM OF DYNAMO]

And here arises a difference and a change of name. All generating machines to this date had been called "*Magneto-electric*" because they used *permanent* steel magnets with which to generate a current by the whirling of the bobbin which we now call an armature. The time came, led to by the improvement of Wilde, in which those steel permanent magnets were no longer used. Then the machine became the "*dynamo-electric*" machine, and leaving off one word, according to our custom, "*dynamo*."

Siemens and Wheatstone almost simultaneously invented so much of the dynamo as was yet incomplete. It has "cores"—the parts that answer to the legs of a horseshoe magnet—of soft iron, sometimes now even of cast iron. These, at starting, possess very little magnetism—practically none at all—yet sufficient to generate a very weak current in the coils, windings, of the armature when it begins to turn. This weak current, passing through the windings of the field magnet, makes these still stronger magnets, and the effect is to evolve a still stronger current in the armature. Soon the full effect is reached. The big iron field magnet, often weighing some thousands of pounds, is then the same as a permanent steel horseshoe magnet, which would hardly be possible at all. One who has watched the installation of a dynamo, knowing that there is nowhere near any ordinary source of electricity, and has seen its armature begin to whirl and hum, and then in a few moments the violet sparklings of the brushes and the evident presence of a powerful current of electricity, is almost justified in the common opinion that the genius of man has devised a machine to *create* something out of nothing. It is true that a *starting* quantity of electricity is required. It exists in almost every piece of iron. Sometimes, to hasten first action, some cells of a galvanic battery are used to pass a current through the coils of the field magnet. After the first use there is always enough magnetism remaining in them during rest or stoppage to make a dynamo efficient after a few moments operation.

[Illustration: PACINOTTI'S RING-ARMATURE DYNAMO.]

This is the dynamo in principle of action. The varieties in construction now in use number scores, perhaps hundreds. Some of them are monsters in size, and evolve a current that is terrific. They are all essentially the same, depending for action upon the laws illustrated in the simplest experiment in induced electricity. One of the best known of the modern machines is Edison's, represented in the picture at the head of this article. In it the field magnet—answering to the horseshoe magnet of the magneto-electric machine—is plainly distinguishable to the unskilled observer. It is not even solid, but is made of several pieces bolted together. Its legs are hollowed at the ends to admit closely the armature which turns there. There are valuable peculiarities in its construction, which, while complying in all respects with the dynamo principle, utilize those principles to the best mechanical advantage. So do others, in other respects that did not occur even to Edison, or were not adopted by him. Probably the modern dynamo is the most efficient, the most accurately measurable, the least wasteful of its power, and the most manageable, of any power-machine so far constructed by man for daily use.

The motor.—This is the twin of the dynamo. In all essentials the two are of the same construction. A difference in the arrangement of the terminals of the wire coils or the wrappings of armature and field magnet, makes of the one a dynamo and of the other a motor. Nevertheless, they are separate studies in electrical science. Practice has brought about modified constructions, as in the case of the dynamo. The differences between the two machines, and their similarities as well, may be explained by a general brief statement.

It is the work of the dynamo to convert mechanical energy into the form of electrical energy. The motor, in turn, changes this electrical energy back again into mechanical energy.

Where the electric light is produced by the dynamo current no motor intervenes. The current is converted into heat and light by merely having an impediment, a restriction, a narrowness, interposed to its free passage on a conducting wire, as heretofore explained, very much as water in a pipe foams and struggles at a narrow place or an obstruction. Where mechanical movements are to be produced by the dynamo current the motor is always the intermediate machine. In the dynamo the armature is rotated by steam power, producing an electrical energy in the form of a powerful current transmitted by a wire. In the motor the armature, in turn, *is rotated by* this current. It is but another instance of that ability to work backwards—to reverse a process—that seems to pervade all machines, and almost all processes. I have mentioned steam power, and, consequently, the necessary burning of coal and expenditure of money in producing the dynamo current. The dynamo and motor are not necessarily economical inventions, but the opposite when the force produced is to be transmitted again, with some loss, into the same mechanical energy that has already been produced by the burning of coal and the making of steam. Across miles of space, and into places where steam would not be possible, the power is invisibly carried. Suggestions of this convenience—stated cases—it is not necessary to cite. The fact is a prominent one, to be noted everywhere.

And it may be made a mechanical economy. The most prominent instance of this is the new utilization of Niagara as a turbine water-power with which to whirl the armatures of gigantic dynamos, using the power thus obtained upon motors, and in the production of light and the transmission of power to neighboring cities.

The discovery of the possibility of transmitting power by a wire, and converting it again into mechanical energy, is a strange story of the human blindness that almost always attends an acuteness, a thinking power, a prescience, that is the characteristic of humanity alone, but which so often stops short of results. This discovery has been attributed to accident alone; the accident of an employé mistaking the uses of wires and fastening their ends in the wrong places. But a French electrician thus describes the occurrence as within his own experience. His name is Hypolyte Fontaine.

But let us first advert to the forgetfulness of the man who really invented the machine that was capable of the opposite action of both dynamo and motor. This was the Italian, Pacinotti. [Footnote: Moses G. Farmer, an American, and celebrated in his day for intelligent electrical researches, is claimed to have made the first reversible motor ever contrived. A small motor made by Farmer in 1847, and embodying the electro-dynamic principle was exhibited at the great exposition at Chicago in 1893. If the genealogy of this machine remains undisputed it fixes the fact that the discovery belongs to this country, and to an American.] He mentioned that his machine could be used either to generate a current of electricity on the application of motive power to its armature, or to produce motive power on connecting it with a source of electricity. Yet it did not occur to him to definitely experiment with two of his machines for the purpose of accomplishing that which in less than twenty years has revolutionized our ideas and practice in transmitted force. He did not suggest that two of his machines could be run together, one as a generator and the other as a motor. He did not think of its advantages with the facilities for it, of his own creation, in his hands.

M. Fontaine states that at the Vienna Exposition of 1873 there was a Gramme machine intended to be operated by a primary battery, to show that the Gramme was capable of being worked by a current, and, as there was also a second machine of the same kind there, of also generating one. These two machines were to demonstrate this range of capacity as *separately worked*, one by power, the other with a battery. There was, then, no intention of coupling them together as late as 1873, with the means at hand and the suggestion almost unavoidable. The dynamo and motor had not occurred to any one. But M. Fontaine states that he failed to get the primary (battery) current in time for the opening, and was troubled by the dilemma. Then the idea occurred to him, as he could do no better, to work one of the machines with a current "deprived," partly stolen, from the other, as a temporary measure. A friend lent him the necessary piece of wire, and he connected the two machines. The machine used as a motor was connected with a pumping apparatus, and when the machine intended as a generator started, and this make-shift, temporarily-stolen current was carried to the acting motor, the action of the last was so much more vigorous than was intended that the water was thrown over the sides of the tank. Fontaine was forced to remedy this excessive action by procuring an additional wire of such length that its resistance permitted the motor to work more mildly and throw less water. This accidentally established the fact of distance, convenience, a revolution in the power of the industrial world. Fontaine states that Gramme had previously told him that he had done the same thing with his machines. The idea was never patented. Neither Pacinotti, who invented the machine originally, nor Gramme, one of the great names of modern electricity, nor this skilled practical electrician, Fontaine, who had charge of the exhibit of the Gramme system at Vienna, considered the fact of the transmission of concentrated power over a thin wire to a great distance as one of value to its inventor or to the industries of mankind. With the motor and the dynamo already made, it was an accident that brought them together after all.

* * * * *

It may be amusing, if not useful, to spend a moment in reviewing of the efforts of men to utilize the power of the electrical current in mechanics before the day of the dynamo and a motor, and while yet the electric light was an infant in the nursery of the laboratory. They knew then, about 1835 to 1870, of the laws of induction as applied to the electro-magnet, or in small machines the generating power, so called, of the magneto-electric arrangement embodied, as a familiar example, in Kidder's medical battery. There is a long list of those inventors, American and European. The first patent issued for an American electro-motor was in 1837, to a man named Thomas Davenport, of Brandon, Vt. He was a man far ahead of his times. He built the first electric railroad ever seen, at Springfield, Mass., in 1835, and considering the means, whose inadequacy is now better understood by any reader of these lines than it then was by the deepest student of electricity, this first railroad was a success. Davenport came as near to solving the problem of an electric motor as was possible without the invention of Pacinotti. Following this there were many patents issued for electro-magnetic motors to persons residing in all parts of the country, north and south. One was made by C. G. Page, of the Smithsonian Institute, in which the motive power consisted in a round rod, acting as a plunger, being pulled into the space where the core would be in an ordinary electro-magnet, and thereby working a crank. [Footnote: The *National Intelligencer*, a prominent Washington newspaper, said with reference to Page's motor "He has shown that before long electro-magnetic action will have dethroned steam and will be the adopted motor," etc. This was an enthusiasm not based upon any fact then known about a machine not even in the line of the present facts of electrodynamics.] A large motor of this kind is alleged, in 1850, to have developed ten horse power. It was actually applied to outdoor experiment as a car-motor on an actual railroad track, and was efficient for several miles. But it carried with it its battery-cells, and they were disarranged and stirred by the jolting, and being made of crockeryware were broken. The chemicals cost much more than fuel for steam, and there could be no economical motive for further experiment. It was a huge toy, as the entire sum of electrical science was until it was made useful first in the one instance of the telegraph, and long after that date the use of the electro-magnet, with a cam to cut off and turn on again the current at proper intervals, which was the one principle of all attempts, was a repeated and invariable failure. That which was wanted and lacking was not known, and was finally discovered and successively developed as has been described.

Electric railroads. — There was an instance of almost simultaneous invention in the case of the first practical electric railroads. S. D. Field, Dr. Siemens, and Thomas A. Edison all applied for patents in 1880. Of these, Field was first in filing, and was awarded patents. The combined dynamo and motor were, of course, the parents of the practical idea. Field's patents covered a motor in or under the car, operated by a current from a stationary source of electricity — of course a dynamo. These first electric roads had the current carried on the rail. They were partially successful, but there was something wrong in the plan, and that something was induction by the earth. Later came, as a remedy for this, the "Trolley" system; the trolley being a small, grooved wheel running upon a current-carrying wire overhead. The question of how best to convey a current to the car-motor is a serious one, doubtless at this moment occupying the attention of highly-trained intelligence everywhere. The motor current is one of high power, and as such intractable; and it is in the character of this current, rather than in methods of insulation, that the remedy for the much-objected-to overhead wire is to be found. It will be remembered that all the phenomena of induction are *unhindered by insulation*.

Aside from the current-carrying problem, the electric road is explainable in all its features upon the theory and practice of the dynamo and motor. It is merely an application of the two machines.

The last is, in usual practice, under the car, and geared to the truck-axle. A more modern mechanical improvement is to make the axle the shaft of the motor armature. When the motor has used the current it passes by most systems into the rail and the ground. By others there is a "metallic circuit"—two wires. Many men whose interest and occupation leads them to a study of such matters know that the use of electricity, instead of steam locomotion, is merely a question of time on all railroads. I have said elsewhere that the actual age of electricity had not yet fully come. It seems to us now that we have attained the end; that there is little more to know or to do. But so have all the generations thought in their day. In the field of electricity there are yet to come practical results of which one may have some foreshadowings in the experiments of men like Tesla, which will make our present times and knowledge seem tame and slow.

Electrolysis.—In all history, fire has been the universal practical solvent. It has been supplanted by the electrical current in some of the most beautiful and useful phenomena of our time. Electrolysis is the name of the process by which fluid chemicals are decomposed by the current.

A familiar early experiment in electrolysis is the decomposition of water—a chemical composed of oxygen and hydrogen, though always thought of and used as a simple, pure fluid. If the poles of a galvanic battery are immersed in water slightly mixed with sulphuric acid to favor electrical action, these poles will become covered with bubbles of gas which presently rise to the surface and pass off. These bubbles are composed of the two constituents of water, the oxygen rising from the positive and the hydrogen from the negative pole. Particles of the substance decomposed are transferred, some to one pole and some to the other; and, therefore, electrolysis is always practiced in a fluid in order that this transference may more readily occur.

The quantity of *electrolyte*—the substance decomposed—that is transferred in a given time is in proportion to the strength of the current. When this electrolyte is composed of many substances a current will act a little on all of them, and the quantity in which the elementary bodies appear at the poles of the current depends upon the quantities of the compounds in the liquid, and on the relative ease with which they yield to the electrical action.

The electrolytic processes are not the mere experiments a brief description of them would indicate, but are among the important processes for the mechanical products of modern times. The extensive nickel-plating that became a permanent fad in this country on the discovery of a special process some years ago, is all done by electrolysis. The silver plating of modern tableware and table cutlery, as beautiful and much less expensive than silver, and the fine finish of the beautiful bronze hardware now used in house-furnishing, are the results of the same process. Some use for it enters into almost every piece of fine machinery, and into the beautifying or preserving of innumerable small articles that are made and used in unlimited quantity.

The process and its principle is general, but there are many details observed in the actual work of electroplating which interest only those engaged. One of the most usual of these is that of making an electrotype. This may mean the making of an exact impression of a medal, coin, or other figure, or a depositing of a coating of the same on any metallic surface. Formerly the faces of the types used in printing were very commonly faced with copper to give them finish and a wearing quality. Even fresh, natural fruits that have been evenly coated with plumbago may be covered with a thin shell of metal. A silver head may be placed on the wood of a walking stick, precisely conforming on the outside to the form of the wood within.

The deposit of metal in the electrotyping process always takes place at the negative pole—the pole by which the current passes out of the fluid into its conductor. This is the "*cathode*." The other is the "*anode*." The "bath," as the fluid in which the process is accomplished is called, for silver, gold or platinum contains one hundred parts of water, ten of potassium cyanide, and one of the cyanide of whichever of those metals is to be deposited. The articles to be plated are suspended in this bath and the battery-power, varying in intensity according to circumstances, is applied. After removal they are buffed and finished. A varying detail is practiced for different metals, and the current now commonly used is from a dynamo.

[Footnote: Among modern modifications of the dynamic current, is its use, modified by proper appliances, for the telegraph and the telephone circuits of cities and the larger towns. Every electric current may now be safely attributed to that source, and from the same circuit and generator all modifications may be produced at once.]

The origin of electrolysis is said to be with Daniell, who noticed the deposit of copper while experimenting with the battery that bears his name. Jacobi, at St. Petersburg, first published a description of the process in 1839. The Elkingtons were the first to actually put the process into commercial practice.

It would be interesting now, were it apropos, to describe the seemingly very ancient processes by which our ancestors gilded, plated, were deceived and deceived others, previous to about 1845. For those things were done, and the genuineness of life has by no means been destroyed by the modern ease with which a precious metal may be deposited upon one utterly base. A contemplation of the moral side of the subject might lead at once to the conclusion that we could now spare one of the least in actual importance of the processes of the all-pervading and wonderful essence that alike makes the lightning-stroke and gilds the plebeian pin that fastens a baby's napkin. But from any other view we could not now dispense with anything electricity does.

General facts.—The names of many of the original investigators of electrical phenomena are perpetuated in the familiar names of electrical measurements. For, notwithstanding its seeming subtlety, there is no force in use, or that has ever been used by men, capable of being so definitely calculated, measured, determined beforehand, as electricity is. As time passes new measurements are adopted and named, some of them being proposed as lately as 1893. An instance of the value of some of these old determinations of a time when all we now know of electrical science was unknown, may be given in what is known as Ohm's Law. Ohm was a native of Erlangen, in Bavaria, and was Professor of Physics at Munich, where he died in 1874. He formulated this Law in 1827, and it was translated into English in 1847. He was recognized at the time, and was given the Copley medal of the Royal Society of London. The Law—for by that distinctive name is it still called, though the name "Ohm," also expresses a unit of measurement—is that *the quantity of current that will pass through a conductor is proportional to the pressure and inversely proportional to the distance.* That is:

Current = Pressure / Resistance.

Transposing the terms of the equation we may get an expression for either of those elements, current, pressure, or resistance, in the terms of the other two. This relation holds true and is accurate in every possible case and condition of practical work. This remarkable precision and definiteness of action has made possible the creation of an extensive school of electrical testing, by which we are not only enabled to make accurate measurement of electrical apparatus and appliances, but also to make determinations in *other* fields by the agency of electricity. When an ocean cable is injured or broken the precise location of the trouble is made *by measuring the electrical resistance of the parts on each side of the injury.*

The magnitudes of measurements of electricity are expressed in the following convenient electrical units:

The VOLT (named from Volta) equals a unit of *pressure* that is equal to one cell of a gravity battery.

The OHM, as a unit of measurement, equals a unit of *resistance* that is equivalent to the resistance of a hundred feet of copper wire the size of a pin.

The AMPÈRE (named from Ampère, 1775-1836, author of a "Collection of Observations on Electro-Dynamics" and other works, and a profound practical investigator) equals a unit of *current* equivalent to the current which one Volt of pressure will produce through one Ohm of wire (or resistance).

The Coulomb (1736—inventor of the means of measuring electricity called the "Torsion balance," and general early investigator) equals a unit of *quantity* of one Ampere flowing for one second.

The Farad (from Faraday, the discoverer of the laws of Induction, see *ante*), equals that unit of *capacity* which is the capacity for holding one Coulomb. Death current.—What is now spoken of as the "Death Current" is one that will instantly overcome the "resistance" of the human, or animal, body. It is a current of from one to two thousand Volts—about the same as that used in maintaining the large arc lights. This question of the killing capacity of the current became officially prominent some years ago, upon the passage by the legislature of the State of New York of a statute requiring the death penalty to be inflicted by means of electricity. The object was to deter evildoers by surrounding the penalty with scientific horror, [Footnote: Hence also the new lingual atrocity, the word "electrocute," derived from "execute" by decapitation and the addition of "electro"] and the idea had its origin in the accidents which formerly occurred much more frequently than now. The "death current" is now almost everywhere, though the care of the men who continually work about "live" wires has grown to be much like that of men who continually handle firearms or explosives, and accidents seldom happen. At first it was apparently difficult for the general public to appreciate the fact that the silent and harmless-looking wires must be avoided. There was suddenly a new and terrific power in common use, and it was as slender, silent and unobtrusive as it was fatal.

Insulation of the hands by the use of rubber gloves, and extreme care, are the means by which those who are called "linemen"—a new industry—protect themselves in their occupation. But there is a new commandment added to the list of those to be memorized by the body-politic. "Do not tread upon, drive over, or touch *any* wire." It may be, and probably is, harmless. But you cannot positively know. [Footnote: It is a common trait of general human nature to refuse to learn save by the hardest of experiences, and so far as the crediting of statements is concerned, to at first believe everything that is not true, and reject most that is. The supernatural, the phenomena of alleged witchcraft and diabolism, and of "luck," "hoodoo," "fate," etc., find ready disciples among those who reject disdainfully the results of the working of natural law. When the railroads were first built across the plains the Indians repeatedly attempted to stop moving trains by holding the ends of a rope stretched across the track in front of the engine, and with results which greatly surprised them When the lines were first constructed in northern Mexico the Mexican peasant could not be induced to refrain from trying personal experiments with the new power, and scores of him were killed before he learned that standing on the track was dangerous. In the United States the era of accidents through indifference to common-looking wires has almost passed, but for some years the fatality was large because people are always governed by appearances connected with *previous* notions, until *new* experiences teach them better.]

INSTRUMENTS OF MEASUREMENT.—Some of the most costly and beautiful of modern scientific instruments are those used in the measurements and determinations of electrical science. There are many forms and varieties for every specific purpose. Electrical measurement has become a department of physical science by itself, and a technical, extensive and varied one. Already the electrical specialist, no more an original experimenter or investigator than the average physician is, has become professional. He makes plans, submits facts, estimates cost, and states results with almost certainty.

ELECTRICITY AS AN INDUSTRY.—Immense factories are now devoted to the manufacture of electrical goods exclusively. Large establishments in cities are filled with them. The installation of the electric plant in a dwelling house is done in the same way, and as regularly, as the plumbing is. Soon there must be still another enlargement, since the heating of houses through a wire, and the kitchen being equipped with cooking utensils whose heat is for each vessel evolved in its own bottom, is inevitable.

The following are some of the facts, in figures, of the business side of electricity in the United States at the present writing. In 1866, about twenty years after the establishment of the telegraph, but with a population of only a little more than half the present, there were 75,686 miles of telegraph wire in use, and 2,520 offices. In 1893 there were 740,000 miles of wire, and more than 20,000 offices. The receipts for the year first named are unknown, but for 1893 they were about $24,000,000. The expenses of the system for the same year were $16,500,000.

The telephone, an industry now about sixteen years old, had in 1893, for the Bell alone, over 200,000 miles of wire on poles, and over 90,000 miles of wire under ground. The instruments were in 15,000 buildings. There were 10,000 employés, and 233,000 subscribers. All companies combined had 441,000 miles of wire. Ninety-two millions of dollars were invested in telephone *fixtures*.

In 1893, the average cost of a telegram was thirty-one and one sixtenths cents, and the average alleged cost of sending the same to the companies was twenty-two and three-tenths cents, leaving a profit of nine and three-tenths cents on every message. It must be remembered that with mail facilities and cheapness that are unrivalled, the telegraph message is always an extraordinary mode of communication; an emergency. These few figures may serve to give the reader a dim idea of the importance to which the most ordinary and general of the branches of electrical industry have grown in the United States.

MEDICAL ELECTRICITY.—For more than fifty years the medical fraternity in regular practice persisted in disregarding all the claims made for the electric current as a therapeutic agent. In earlier times it was supposed to have a value that supplanted all other medical agencies. Franklin seems to have been one of the earliest experimenters in this line, and to have been successful in many instances where his brief spark from the only sources of the current then known were applicable to the case. The medical department of the science then fell into the hands of charlatans, and there is a natural disposition to deal in the wonderful, the miraculous or semi-miraculous, in the cure of disease. Divested of the wonder-idea through a wider study and greater knowledge of actual facts, electricity has again come forward as a curative agent in the last ten years. Instruction in its management in disease is included in the curriculum of almost every medical school, and most physicians now own an outfit, more or less extensive, for use in ordinary practice. To decry and utterly condemn is no longer the custom of the steady-going physician, the ethics of whose cloth had been for centuries to condemn all that interfered with the use of drugs, and everything whose action could not be understood by the examples of common experience, and without special study outside the lines of medical knowledge as prescribed.

Perhaps the developments based upon the discoveries of Faraday have had much to do with the adoption of electricity as a curative agent. The current usually used is the Faradic; the induced alternate current from an induction coil. This is, indeed, the current most useful in the majority of the nervous derangements in the treatment of which the current is of acknowledged utility.

In surgery the advance is still greater. "Galvano-cautery" is the incandescent light precisely; the white-hot wire being used to cut off, or burn off, and cauterize at the same time, excrescences and growths that could not be easily reached by other means than a tube and a small loop of platinum wire. A little incandescent lamp with a bulb no bigger than a pea is used to light up and explore cavities, and this advance alone, purely mechanical and outside of medical science, is of immense importance in the saving of life and the avoidance of human suffering.

It may be added that there is nothing magical, or by the touch, or mysterious, in the treatment of disease by the electrical current. The results depend upon intelligent applications, based upon reason and experience, a varied treatment for varying cases. Nor is it a remedy to be applied by the patient himself more than any other is. On the contrary, he may do himself great injury. The pills, potions, powders and patent medicines made to be taken indiscriminately, and which he more or less understands, may be still harmful yet much safer. Even the application of one or the other of the two poles with reference to the course of a nerve, may result in injury instead of good.

INCOMPLETE POSSIBILITIES.—There are at least two things greatly desired by mankind in the field of electrical science and not yet attained. One of these, that may now be dismissed with a word, is the resolving of the latent energy of, say a ton of coal, into electrical energy without the use of the steam engine; without the intervention of any machine. For electricity is not manufactured; not created by men in any case. It exists, and is merely gathered, in a measure and to a certain extent confined and controlled, and sent out as a *concentrated form of energy* on its various errands. Should a means for the concentration of this universally diffused energy be found whereby it could be made to gather, by the new arrangement of some natural law such as places it in enormous quantities in the thundercloud, a revolution that would permeate and visibly change all the affairs of men would take place, since the industrial world is not a thing apart, but affects all men, and all institutions, and all thought.

The other desideratum, more reasonable apparently, yet far from present accomplishment, is a means of storing and carrying a supply of electricity when it has been gathered by the means now used, or by any means.

THE STORAGE BATTERY is an attempt in this last direction. The name is misleading, since even in this attempt electricity is in no sense "stored," but a chemical action producing a current takes place in the machine. The arrangement is in its infancy. Instances occur in which, under given circumstances, it is more or less efficient, and has been improved into greater efficiency. But many difficulties intervene, one of which is the great weight of the appliances used, and another, considerable cost. The term "storage battery" is now infrequently used, and the name "secondary" battery is usually substituted. The principle of its action is the decomposing of combined chemicals by the action of a current applied from a stationary generator or dynamo, and that these chemicals again unite as soon as they are allowed to do so by the completing of a circuit, *and in recombining give off nearly as much electricity as was first used in separating them.* The action of the secondary, "storage," battery, once charged, is like that of a primary battery. The current is produced by chemical action. Two metals outside of the solution contained in a primary battery cell, but under differing physical conditions from each other, will yield a current. A piece of polished iron and a piece of rusty iron, connected by a wire, will yield a small current. Rusty lead, so to speak, so connected with bright lead, has a high electromotive force. Oxygen makes lead rusty, and hydrogen makes it bright. Oxygen and hydrogen are the two gases cast off when water is subjected to a current. (See *ante* under *Electrolysis*) So Augustin Planté, the inventor of as much as we yet have of what is called a storage or secondary battery, suspended two plates of lead in water, and when a current of electricity was passed through it hydrogen was thrown off at one plate, making it bright, and oxygen at the other plate, peroxydizing its surface. When the current was removed the altered plates, connected by a wire, would send off a current which was in the opposite direction from the first, and this would continue until the plates were again in their original condition. This is the principle and mode of action of the storage battery. So far it has assumed many forms. Scores of modifications have been invented and patented. The leaden plates have taken a variety of forms, yet have remained leaden plates, one cleaned and the other fouled by the electrolytic action of a current, and giving off an almost equivalent current again by the return process. The arrangement endures for several repetitions of the process, but is finally expensive and always inconvenient. The secondary battery, in its infancy, as stated, presents now much the same obstacles to commercial use the galvanic, or primary, battery did before the induced current had become the servant of man.

CHAPTER IV.

ELECTRICAL INVENTION IN THE UNITED STATES.

A list of the electrical inventors of this country would be very long. Many of the names are, in the mass and number of inventions, almost lost. It happens that many of the practical applications described in this volume, indeed most of them, are the work of citizens of this country.

In previous chapters I have referred briefly to Franklin, Morse, Field, and others. These men have left names that, without question, may be regarded as permanent. Their chiefest distinguishing trait was originality of idea, and each one of them is a lesson to the American boy. In a sense the greatest of all these, and in the same sense, the greatest American, was Benjamin Franklin. A sketch of his career has been given, but to that may be added the following: He had arrived at conclusions that were vast in scope and startling in result by applying the reasoning faculty upon observations of phenomena that had been recurring since the world was made, and had been misunderstood from the beginning. He used the simplest means. His experiment was in a different way daily performed for him by nature. He was philosophically daring, indifferently a tinker with nature's terrific machinery; a knocker at the door of an august temple that men were never known to have entered; a mortal who smiled in the face of inscrutable and awful mystery, and who defied the lightning in a sense not merely moral. [Footnote: Professor Richmann, of St. Petersburg, was instantly killed by lightning while repeating Franklin's experiment.]

His genius lay in a power of swift inductive reasoning. His common sense and his sense of humor never forsook him. He uttered keen apothegms that have lived like those of Solon. He was a philosopher like Diogenes, lacking the bitterness. He wrote the "Busy-Body," and annually made the plebeian and celebrated "Almanac," and the "Ephemera" that were not ephemeral, and is the author of the story of "The Whistle," that everybody knows, and everybody reads with shamefacedness because it is a brief chapter out of his own history.

He was apparently an adept in the art of caring for himself, one of the most successful worldings of his time, yet he wrote, thought, toiled incessantly, for his fellow men. He had little education obtained as it is supposed an education must be obtained. He was commonplace. No one has ever told of his "silver tongue," or remembered a brilliant after-dinner speech that he has made. Yet he finally stood before mankind the companion of princes, the darling of splendid women, covered with the laurels of a brilliant scientific renown. But he was a printer, a tinkerer with stoves, the inventor of the lightning rod, the man who had spent one-half his life in teaching apprentices, such as he himself had been when his jealous and common-minded brother had whipped him, that "time is money," that "credit is money"—which is the most prominent fact in the commercial world of 1895—and that honor and self-respect are better than wealth, pleasure, or any other good.

Yet clear, keen, cold and inductive as was Franklin's mind, no vision reached him, in the moment of that triumph when he felt the lightning tingling in his fingers from a hempen string, of those wonders which were to come. He knew absolutely nothing of that necromancy through which others of his countrymen were to girdle the world with a common intelligence, and yet others were to use in sprinkling night with clusters as innumerable and mysterious as the higher stars.

The story of the Morse telegraph has been repeatedly told, and I have briefly sketched it in connection with the subject of the telegraph. But, unlike the original, scientifically lonely and independent Franklin, Morse had the best assistance of his times in the persons of men more skilled than himself and almost as persistent. The chief of these was Alfred Vail, a name until lately almost unknown to scientific fame, who eliminated the clumsy crudities of Morse's conception, remade his instruments, and was the inventor of that renowned alphabet which spells without letters or writing or types, that may be seen or heard or felt or tasted, that is adapted to any language and to all conditions, and that performs to this day, and shall to all time, the miracle of causing the inane rattle of pieces of metal against each other to speak to even a careless listener the exact thoughts of one a thousand miles away.

Another of the men who might be appropriately included in any comprehensive list of aiders and abettors of the present telegraph system were Leonard D. Gale, then Professor of Chemistry in the University of New York, and Professor Joseph Henry, who had made, and was apparently indifferent to the importance of it because there was no alphabet to use it with, the first electric telegraph ever constructed to be read, or used, *by sound*. Last, though hardly least if all facts are understood, might be included a skillful youth named William Baxter, afterwards known as the inventor of the "Baxter Engine," who, shut in a room with Vail in a machine shop in New Jersey, made in conjunction with the author of the alphabet the first telegraphic instrument that, with Henry's magnet and battery cells, sent across space the first message ever read by a person who did not know what the words of the message would say or mean until they had been received.

After the telegraph the state of electrical knowledge was for a long time such that electrical invention was in a sense impossible. The renowned exploit of Field was not an invention, but a heroic and successful extension of the scope and usefulness of an invention. But thought was not idle, and filled the interval with preparations for final achievements unequaled in the history of science. Two of these results are the electric light and the telephone. For the various "candles," such as that of Jablochkoff, exhibited at Paris in 1870, only served to stimulate investigation of the alluring possibilities of the subject. The details of these great inventions are better known than those of any others. The telegraph and the newspaper reporter had come upon the field as established institutions. Every process and progress was a piece of news of intense interest. When the light glowed in its bulb and sparkled and flashed at the junction points of its chocolate-colored sticks it had been confidently expected. There was little surprise. The practical light of the world was considered probable, profitable, and absolutely sure. The real story will never be told. The thoughts, which phrase may also include the inevitable disappointments of the inventor, are never written down by him. That variety of brain which, with a few great exceptions, was not known until modern, very recent times, which does not speculate, contrive, imagine only, but also reduces all ideas to *commercial* form, has yet to have its analysis and its historian, for it is to all intents a new phase of the evolution of mind.

[Illustration: THOMAS A. EDISON.]

A typical example of this class of intellect is Mr. Thomas A. Edison. It may be doubted if such a man could, in the qualities that make him remarkable, be the product of any other country than ours. In common with nearly all those who have left a deep impression upon our country, Edison was the child of that hackneyed "respectable poverty" which here is a different condition from that existing all over Europe, where the phrase was coined. There, the phrase, and the condition it describes, mean a dull content, an incapacity to rise, a happy indifference to all other conditions, a dullness that does not desire to learn, to change, to think. To respectable poverty in other civilizations there are strong local associations like those of a cat, not arising to the dignity of love of country. In the United States, without a word, without argument or question, a young man becomes a pioneer—not necessarily one of locality or physical newness, but a pioneer in mind—in creed, politics, business—in the boundless domain of hope and endeavor. In America no man is as his father was except in physical traits. No man there is a volunteer soldier fighting his country's battles except from a conviction that he ought to be. A man is an inventor, a politician, a writer, first because he knows that valuable changes are possible, and, second, because he can make such changes profitable to himself. It is the great realm of immutable steadfastness combined with constant change; unique among the nations.

Edison never had more than two months regular schooling in his entire boyhood. There is, therefore, nothing trained, "regular," technical, about him. If there had been it is probable that we might never have heard of him. He is one of the innumerable standing arguments against the old system advocated by everybody's father, and especially by the older fathers of the church, and which meant that every man and woman was practically cut by the same pattern, or cast in the same general mould, and was to be fitted for a certain notch by training alone. No more than thirty years ago the note of preparation for the grooves of life was constantly sounded. Natural aptitude, "bent," inclination, were disregarded. The maxim concocted by some envious dull man that "genius is only another name for industry," was constantly quoted and believed.

But Edison's mother had been trained, practically, as an instructor of youth. He had hints from her in the technical portions of a boy's primary training. He is not an ignorant man, but, on the contrary, a very highly educated one. But it is an education he has constructed for himself out of his aptitudes, as all other actual educations have really been. When he was ten years old he had read standard works, and at twelve is stated to have struggled, ineffectually perhaps, with Newton's *Principia*. At that age he became a train-boy on the Grand Trunk railroad for the purpose of earning his living; only another way of pioneering and getting what was to be got by personal endeavor. While in that business he edited and printed a little newspaper; not to please an amateurish love of the beautiful art of printing, but for profit. He was selling papers, and he wanted one of his own to sell because then he would get more out of it in a small way. He never afterwards showed any inclination toward journalism, and did not become a reporter or correspondent, or start a rural daily. While he was a train-boy, enjoying every opportunity for absorbing a knowledge of human nature, and of finally becoming a passenger conductor or a locomotive engineer, something called his attention to the telegraph as a promoter of business, as a great and useful institution, and he resolved to become an "operator." This was his electrical beginning. Yet before he took this step he was accused of a proclivity toward extraordinary things. In the old "caboose" where he edited, set up, and printed his newspaper he had established a small chemical laboratory, and out of these chemicals there is said to have been jolted one day an accident which caused him some unpopularity with the railroad people. He was all the time a business man. He employed four boy helpers in his news and publishing business. It took him a long time to learn the telegraph business under the circumstances, and when he was at last installed on a "plug" circuit he began at once to do unusual things with the current and its machines and appliances. This is what he tells of his first electrical invention.

There was an operator at one end of the circuit who was so swift that Edison and his companion could not "take" fast enough to keep up with him. He found two old Morse registers—the machines that printed with a steel point the dots and dashes on a paper slip wound off of a reel. These he arranged in such a way that the message written, or indented, on them by the first instrument were given to him by the second instrument at any desired rate of speed or slowness.

This gave to him and his friend time to catch up. This, in Morse's time, would have been thought an achievement. Edison seems to regard it as a joke. There was no time for prolonged experiment. It was an emergency, and the idea must necessarily have been supplemented by a quick mechanical skill.

It was this same automatic recorder, the idea embodied in it, that by thought and logical deduction afterwards produced that wonderful automaton, the phonograph. He rigged a hasty instrument that was based upon the idea that if the indentations made in a slip of paper could be made to repeat the ticking sound of the instrument, similar indentations made by a point on a diaphragm that was moved by the *voice* might be made to repeat the voice. His rude first instrument gave back a sound vaguely resembling the single word first shouted into it and supposed to be indented on a slip of paper, and this was enough to stimulate further effort. He finally made drawings and took them to a machinist whom he knew, afterwards one of his assistants, who laughed at the idea but made the model. Previously he bet a friend a barrel of apples that he could do it. When the model was finished he arranged a piece of tin foil and talked into it, and when it gave back a distinct sound the machinist was frightened, and Edison won his barrel of apples, "which," he says, "I was very glad to get."

The "Wizard" is a man evidently pertaining to the class of human eccentrics who excite the interest of their fellow-men "to see what they will do next," but without any idea of the final value of that which may come by what seems to them to be mere unbalanced oddity. Such people are invariably misunderstood until they succeed. When he invented the automatic repeating telegraph he was discharged, and walked from Decatur to Nashville, 150 miles, with only a dollar or two as his entire possessions. With a pass thence to Louisville, he and a friend arrived at that place in a snowstorm, and clad in linen "dusters." This does not seem scientific or professor-like, but it has not hindered; possibly it has immensely helped. It reminds one of the Franklinic episodes when remembered in connection with future scientific renown and the court of France.

One of the secrets of Edison's great success is the ease with which he concentrates his mind. He is said to possess the faculty of leaving one thing and taking up another whenever he wills. He even carries on in his mind various trains of thought at the same time. The operations of his brain are imitated in his daily conduct, which is direct and simple in all respects. He is never happier than when engaged in the most absorbing and exacting mental toil. He dresses in a machinist's clothes when thus employed in his laboratory, and was long accustomed to work continuously for as long as he was so inclined without regard to regularity, or meals, or day or night. He is willing to eat his food from a bench that is littered with filings, chips and tools. To relieve strain and take a moment's recreation he is known to have bought a "cottage" organ and taught himself to play it, and to go to it in the middle of the night and grind out tunes for relaxation. He has a working library containing several thousand books. He pores over these volumes to inform himself upon some pressing idea, and does so in the midst of his work. No man could have made some of his inventions unaided by technical science and a knowledge of the results of the investigations of many others, and it has often been wondered how a man not technically educated could have seemed so well to know. There was a mistake. He *is* educated; a scientific investigator of remarkable attainments.

In thinking of the inventions of Edison and their value, a dozen of the first class, that would each one have satisfied the ambition or taken the time of an ordinary man, can be named. The mimeograph and the electric pen are minor. Then there are the stock printer, the automatic repeating telegraph, quadruplex telegraphy, the phonoplex, the ore-milling process, the railway telegraph, the electric engine, the phonograph. Some of these inventions seem, in the glow of his incandescent light, or with one's ear to the tube of the telephone he improved in its most essential part, to be too small for Edison. But nothing was too small for Franklin, or for the boy who played idly with the lid of his mother's tea-kettle and almost invented the steam-engine of today, or for Hero of Alexandria, who dreamed a thousand years before its time of the power that was to come. So was Henry's first electric telegraph the merest toy, and his electro-magnet was supported upon a pile of books, his signal bell was that with which one calls a servant, and his idea was a mere experiment without result. There was a boy Edison needed there then, whose toys reap fortunes and light, and enlighten, the world. The electric pen was in its day immensely useful in the business world, because it was the application of the stencil to ordinary manuscript, and caused the making of hundreds of copies upon the stencil idea, and with a printer's roller instead of a brush. The mimeograph was the same idea in a totally different form. It was writing upon a tablet that is like a bastard-file, with a steel-pointed stylus. Each slight projection makes a hole in the paper, and then the stencil idea begins again.

Something has been previously said of the difficulties attending the making of the filament for the incandescent light. It is a little thing, smaller than a thread, frail, delicate, sealed in a bulb almost absolutely exhausted of air, smooth without a flaw, of absolutely even caliber from end to end. The world was searched for substances out of which to make it, and experiments were endlessly and tediously tried; all for this one little part of a great invention, which, like all other inventions, would be valueless in the want of a single little part.

There are hundreds, an unknown number, of inventions in electricity in this country whose authors are unknown, and will never be known to the general public. The patent office shows many thousands of such in the aggregate. Many useful improvements in the telephone alone have come under the eye of every casual reader of the newspapers. These are now locked up from the world, with many other patented changes in existing machines, because of the great expense attending their substitution for those arrangements now in use.

All the principles—the principles that, finally demonstrated, become laws—upon which electrical invention is based, are old. It seems impossible, during the entire era of modern thought, to have found a new trait, a development, a hitherto unsuspected quality. Tesla, in some of his most wonderful experiments, seems almost to have touched the boundaries of an unexplored realm, yet not quite, not yet, and most likely absolute discovery can no farther go. To play upon those known laws—to twist them to new utilities and give them new developments—has been the work of the creators of all the modern electrical miracles. There is scarcely a field in which men work in which the results are not more apparent, yet all we have, and undoubtedly most we shall ever have, of electricity we shall continue to owe to the infant period of the science.

It may be truthfully claimed that most of these extraordinary applications of electricity have been made by American inventors. Wherever there is steam, on sea or land, there, intimately associated with American management, will be found the dynamic current and all its uses. The science of explosive destruction has almost entirely changed, and with a most extraordinary result. But one of the factors of this change has been the electric current, a something primarily having nothing to do with guns, ships or sailing. The modern man-of-war, beginning with those of our own navy, is lighted by the electric light, signalled and controlled by the current, and her ponderous guns are loaded, fired, and even *sighted* by the same means. Her officers are a corps of electrical experts. A large part of her crew are trained to manipulate wires instead of ropes, and her total efficiency is perhaps three times what it would be with the same tonnage under the old régime. There is a new sea life and sea science, born full grown within ten years from a service encrusted with traditions like barnacles, and that could not have come by any other agency. A big gun is no longer merely that, but also an electrical machine, often with machinery as complicated as that of a chronometer and much more mysterious in operation.

I have said that the huge piece was even sighted by electricity. There is really nothing strange in the statement, though it may read like a fairy tale or a metaphor to whoever has never had his attention called to the subject. In a small way, with the name of its inventor almost unknown except to his messmates, it is one of the most wonderful, and one of the simplest, of the modern miracles. As a mere instance of the wide extent of modern ideas of utility, and of the possibilities of application of the laws that were discovered and formulated by those whose names the units of electrical measurements bear, it may be briefly stated how a group of gunners may work behind an iron breastwork, and never see the enemy's hull, and yet aim at him with a hundred times the accuracy possible in the day of the *Old Ironsides* and the *Guerriere*.

And first it may be stated that the *range-finder* is largely a measure of mere economy. A two-million-dollar cruiser is not sailed, or lost, as a mere pastime. Whoever aims best will win the fight. Ten years ago the way of finding distance, or range, which is the same thing, was experimental. If a costly shot was fired over the enemy the next one was fired lower, and possibly between the two the range might be got, both vessels meantime changing positions and range. To change this, to either injure an antagonist quickly or get away, the "range-finder" was invented, as a matter not of business profit, by Lieutenant Bradley A. Fiske, of the U. S. Navy, in 1889. It has its reason in the familiar mathematical proposition that if two angles and one side of a triangle are known, the other sides of the triangle are easily found. That is, that it can be determined how far it is to a distant object without going to it. But Fiske's range-finder makes no mathematical calculations, nor requires them to be made, and is automatic. A base line permanently fixed on the ship is the one side of a triangle required. The distance of the object to be hit is determined by its being the apex of an imaginary triangle, and at each of the other angles, at the two ends of the base line, is fixed a spyglass. These are directed at the object.

So far electricity has had nothing to do with the arrangement, but now it enters as the factor without which the device could have no adaptation. As the telescopes are turned to bear upon the target they move upon slides or wires bent into an arc, and these carry an electric current. The difference in length of the slide passed over in turning the telescopes upon the object causes a greater or less resistance to the current, precisely as a short wire carries a current more easily; with less "resistance;" than a long one. A contrivance for measuring the current, amounting to the same thing that other instruments do of the same class that are used every day, allows of this resistance being measured and read, not now in units of electricity, but *in distance to the apex of the triangle where the target is*; in yards. The man at each telescope has only to keep it pointed at the target as it moves, or as the vessel moves which wishes to hit it. And now even the telephone enters into the arrangement. Elsewhere in the ship another man may stand with the transmitter at his ear. He will hear a buzzing sound until the telescopes stop moving, and at the same time there will be under his eye a pointer moving over a graduated scale. The instant the sound ceases he reads the range denoted by the index and scale. The information is then conveyed in any desired way to the men at the guns; these, of course, being aimed by a scale corresponding to that under the eye of the man at the telephone. The plan is not here detailed as technical information valuable to the casual reader, but as showing the wide range of electrical applications in fields where possible usefulness would not have been so much as suspected a few years ago. The same gentleman, Lieut. Fiske, is also the author of ingenious electrical appliances for the working of those immense gun-carriages that have grown too big for men to move, and for the hoisting into their cavernous breeches of shot and shell. The men who work these guns now do not need to see the enemy, even through the porthole or the embrasure. They can attend strictly to the business of loading and firing, assisted by machines nearly or quite automatic, and can cant and lay the piece by an index, and fire with an electric lanyard. The genius of science has taken the throne vacated by the goddess of glory. The sailor has gone, and the expert mechanician has taken his place. The tar and his training have given way to the register, the gauge and the electrometer. The big black guns are no longer run backward amid shouts and flying splinters, and rammed by men stripped to the waist and shrouded in the smoke of the last discharge, but swing their long and tapering muzzles to and fro out of steel casemates, and tilt their ponderous breeches like huge grotesque animals lying down. The grim machinery of naval battle is moved by invisible hands, and its enormous weight is swayed and tilted by a concealed and silent wire.

This strange slave, that toils unmoved in the din of battle, has been reduced to domestic servitude of the plainest character. The demonstrations made of cooking by electricity at the great fair of 1893 leave that service possible in the future without any question. Electrical ovens, models of neatness, convenience and *coolness*, were shown at work. They were made of wood, lined with asbestos, and were lighted inside with an incandescent lamp. The degree of temperature was shown by a thermometer, and mica doors rendered the baking or roasting visible. There could be no question of too much heat on one side and too little on another, because switches placed at different points allowed of a cutting off, or a turning on, whenever needed. Laundry irons had an insulated pliable connection attached, so that heat was high and constant at the bottom of the iron and not elsewhere. There were all the appliances necessary for the broiling of steaks, the making of coffee and the baking of cakes, and the same mystery, which is no longer a mystery, pervaded it all. Woman is also to become an electrician, at least empirically, and in time soon to come will understand her voltage and her Ampères as she now does her drafts and dampers and the quality of her fuel.

It is a practical fact that chickens are hatched by the thousand by the electrical current, and that men have discovered more than nature knew about the period of incubation, and have reduced it by electricity from twenty-one to nineteen days. The proverb about the value of the time of the incubating hen has passed into antiquity with all things else in the presence of electrical science.

Whenever an American mechanician, a manufacturer or an inventor, is confronted by a difficulty otherwise insolvable he turns to electricity. Its laws and qualities are few. They seem now to be nearly all known, but the great curiosity of modern times is the almost infinite number of applications which these laws and qualities may be made to serve. One may turn at a single glance from the loading and firing of naval guns to the hatching of chickens and the cooking of chocolate by precisely the same means, silently used in the same way. Most of these applications, and all the most extraordinary ones, are of American origin. Their inventors are largely unknown. There is no attempt made here to more than suggest the possibilities of the near future by a glimpse of the present. The generation that is rising, the boy who is ten years old, should easily know more of electrical science than Franklin did. There are certain primal laws by which all explanations of all that now is, and most probably of almost all that is to come so far as principles go, may be readily understood, and these I have endeavored, in this and preceding chapters, to explain.

There are in the United States new applications of electricity literally every day. Before the written page is printed some startling application is likely to be made that gives to that page at once an incompleteness it is impossible to guard against or avoid. There is a strong inclination to prophesy; to tell of that which is to come; to picture the warmed and illuminated future, smokeless and odorless, and the homes in which the children of the near future shall be reared. Some of those few apprehended things, suggested as being possible or desirable in these chapters, have been since done and the author has seen them. This American facility of electrical invention has one great cause, one specific reason for its fruitfulness. It is because so many acute minds have mastered the simple laws of electrical action. This knowledge not only fosters intelligent and fruitful experiment but it prevents the doing of foolish things. No man who has acquired a knowledge of mechanical forces, who understands at least that great law that for all force exerted there is exacted an equivalent, ever dreams upon the folly of the perpetual motion. In like manner does a knowledge, purely theoretical, of the laws of electricity prevent that waste of time in gropings and dreams of which the story of science and the long human struggle in all ages and in all departments is full.

Finally, I would, if possible dispell all ideas of strangeness and mystery and semi-miracle as connected with electrical phenomena. There is no mystery; above all, there is no caprice. There are, in electricity and in all other departments of science, still many things undiscovered. It is certain that causes lead far back into that realm which is beyond present human investigation. *Force* has innumerable manifestations that are visible, that are understood, that are controlled. Its *origin* is behind the veil. A thousand branching threads of argument may be taken up and woven into the single strand that leads into the unknown. Out of the thought that is born of things has already arisen a new conception of the universe, and of the Eternal Mind who is its master. Among these things, these daily manifestations of a seeming mystery, the most splendid are the phenomena of electricity. They court the human understanding and offer a continual challenge to that faculty which alone distinguishes humanity from the beasts. The assistance given in the preceding pages toward a clear understanding of the reason why, so far as known, is perhaps inadequate, but is an attempt offered for what of interest or value may be found.

www.ingramcontent.com/pod-product-compliance
Lightning Source LLC
Chambersburg PA
CBHW031419210526
45464CB00005B/1955